The ENCON Office Building. A view of the west facade (at left) and the south facade showing the solar collector and the deep louvers.

SOLAR ENERGY

Technology
and
Applications

Revised Edition

J. Richard Williams, Ph.D.

Associate Dean for Research
College of Engineering
Georgia Institute of Technology
Atlanta, Georgia

ANN ARBOR SCIENCE
PUBLISHERS INC
P.O. BOX 1425 • ANN ARBOR, MICH. 48106

Revised Edition © 1977
Library of Congress Catalog Card Number 77-73635
ISBN 0-250-40194-0

Copyright © 1974 by Ann Arbor Science Publishers, Inc.
230 Collingwood, P.O. Box 1425, Ann Arbor, Michigan 48106
ISBN 0-250-40064-2

Manufactured in the United States of America

PREFACE TO THE SECOND EDITION

Solar energy represents the only totally nonpolluting inexhaustible energy resource that can be utilized economically to supply Man's energy needs for all time. Recent large price increases and reduced availability of fossil fuels, such as evidenced in the energy crisis of the winter of 1977, and public concern about the safety of nuclear reactors have led to a surge of interest in using the power of the sun.

Solar Energy Technology and Applications introduces the various techniques for utilizing solar energy and brings you up to date on work to the present time on the broad spectrum of solar energy systems. Succinctly written for the technical person, it is readily understood by anyone with no background in the field. It is also recommended as a supplementary text for energy-related courses, and includes information on how to install solar systems for space heating and hot water.

The Second Edition includes more than 40 new pages of specific installations plus 100 or more smaller additions and updates throughout the text.

J. Richard Williams, 1977

ACKNOWLEDGMENTS

I would like to express my appreciation for the support of my research by the National Aeronautics and Space Administration from 1967 to 1973, which introduced me to the exciting new field of solar energy. The guidance and assistance of Robert Ragsdale and George Kaplan of the NASA Lewis Research Center during these years is gratefullly acknowledged.

J. Richard Williams

CONTENTS

Chapter 1

SOLAR ENERGY: ITS TIME HAS COME

In 1970 the total energy consumption in the United States was about 65 $\times$ 10^{15} BTU, which is equal to the energy of sunlight received by 4300 square miles of land, or only 0.15% of the land area of the continental U.S.[1] Even if this energy were utilized with an efficiency of only 10%, the total energy needs of the U.S. could be supplied by solar collectors covering only 1.5% of the land area, and this energy would be supplied without any environmental pollution. With the same 10% utilization efficiency, about 4% of the land area could supply all the energy needs in the year 2000. By comparison, at present 15% of the U.S. land area is used for growing farm crops.[2] For some applications, such as heating water and space heating for buildings, the utilization efficiency can be much greater than 10%, and the collectors can be located on vertical walls and rooftops of buildings, so the 4% estimate represents an upper limit and actual land area requirements can be considerably smaller.

As a practical matter, even though sunlight can provide all our energy needs without pollution, in the foreseeable future solar energy will not provide all or even most of this energy. Over the past century fossil fuels have provided most of our energy because energy from fossil fuels has usually been cheaper and more convenient than energy from available alternative energy sources, and until recently environmental pollution has been of little concern. The construction of large nuclear electric generating plants is presently underway, and nuclear power will play an increasingly important role; so in the coming decades a variety of energy sources

1

will supply the U.S. energy needs, and solar energy will only be utilized when it is competitive with alternative energy sources.

Over the past few years energy forecasts[3-12] have been made which predict large increases in the consumption of oil and coal as well as a rapid increase in nuclear generation. However, these forecasts predict that the domestic production of oil would not be sufficient to keep pace with demand, so large increases in oil imports would be necessary. The recent rapid escalation of the cost of foreign crude oil has cast doubt on the ability of the U.S. to supplement its energy needs from foreign imports, so the President has urged that the U.S. become self-sufficient in its energy supplies. This will require developing additional domestic energy resources. Solar energy, which so far has seen insignificant use in the U.S., can be utilized to make a significant impact as a new energy resource over the next few years. The most immediate large-scale applications would be for the heating and cooling of buildings, heating water, and supplying heat for industrial and agricultural drying operations. Over the longer term, solar energy can also be used for pollutionless electric power generation.

The NSF/NASA Solar Energy Panel[13] identified three broad applications as "most promising from technical, economic, and energy quantity standpoints. These are: (1) the heating and cooling of residential and commercial buildings, (2) the chemical and biological conversion of organic materials to liquid, solid, and gaseous fuels, and (3) the generation of electricity." It also reported that "solar energy can be developed to meet sizable portions of the nation's future energy needs." Energy for space heating, air conditioning, and water heating for buildings presently accounts for about 25% of the total energy consumption in the U.S.,[14] and virtually all this energy is supplied by the combustion of high-quality fossil fuels. Sunlight can provide about half this energy at a cost competitive with fossil fuels, thus reducing total consumption of fossil fuels by more than 12%.[15-16] In the future sunlight can also be utilized directly for electric power generation and for producing fuels to replace the fossil fuels now being used as an energy source. Coal, oil, and gas are nonrenewable resources of great value for producing fabrics, plastics and other materials, and probably should not be burned once economical alternative fuels become available.

More was accomplished between 1974 – 1976 toward the development of practical, economical solar energy systems than had been accomplished prior to 1974. The reason for this rapid advance of solar energy technology is the increasing interest in the field and the tremendous increase in financial support for solar energy research and development throughout the world. In the United States the Energy Research and Development Agency has been created and is supporting solar energy research at a funding level of over $200 million per year.

Major technological breakthroughs in the solar energy field during the last few years include the development of commercially-available selective coatings which improve the performance of flat plate solar collectors operating at the higher temperatures required for air conditioning, and the introduction of new types of solar collectors into the marketplace. Also solar energy data are becoming more readily available in a format which can be utilized to determine the performance of proposed solar energy systems.

Grants from the federal government have supported the construction of a variety of solar homes and buildings, as well as the retrofit of existing facilities for solar heating and cooling. For example, in 1976 the Department of Housing and Urban Development supported the installation of solar systems for 143 single- and multi-family units. The total number of solar homes built and under construction in the U.S. has escalated rapidly and as of 1977 there are at least 1000 solar homes under construction or completed, both with federal and private support. The demand for solar collectors and other solar systems components has stimulated manufacturing companies to make solar hardware available. Only a few years ago commercially manufactured flat plate solar collectors, differential thermostats, pumps for home solar water heating applications, and solar air conditioning units were difficult to obtain within the U.S. Now these products are readily available from a variety of manufacturers.

Another factor which has become apparent is that optimistic estimates of a few years ago as to the cost of solar systems could not be realized; whereas predictions of solar collector costs were in the neighborhood of $4/square foot, commercially available solar collectors today run approximately $10/square foot and sometimes even higher. One reason for this cost escalation is inflation. The escalation in the cost of solar hardware has been accompanied by an even greater escalation in energy costs, so that the economics of solar heating and cooling today are generally even more favorable than ever. At the present time the cost of a home solar heating system, which also provides domestic hot water, runs about $10,000 more than conventional heating systems. If solar heating, air conditioning and domestic hot water are desired, the cost increases to around $15,000. The payback time (the length of time required to save enough money to pay for the solar system) depends on the cost of alternative fuels. For solar heating, or heating and cooling, systems for homes and buildings the payback time is usually in the range of 15 to 20 years, providing that natural gas is not available as an alternative fuel. Natural gas, where it is available, is still more economical than solar energy, but prices are rising and reserves are dwindling. In the next few years the price of natural gas can be expected to escalate to that of other alternative fuels, or its availability will be even less than today.

There are at present over 100 corporations scattered throughout the U.S. offering solar hot water heaters for sale. Solar-heated domestic hot water is still the single most economically advantageous application of solar energy, since the payback time for such a system is typically between 3 and 5 years. Home solar water heaters range in price between $500 and $2000, depending on the type of system employed. The less expensive solar water heaters tend to be passive, and such natural-circulation water heaters are only applicable to southern climates where freezing is less of a problem. Also these passive systems generally require the location of a hot water tank either on the roof or in the attic of the home, so that the tank is higher than the collectors. In order to reduce the cost and aesthetic problems associated with installing this type of solar water heater on a house, some companies are offering units which sit on the ground, complete with solar collectors and a tank situated above the collector. These units are useful only when there is a generally unshaded location available close to the house. Pumped systems cost over $1000, but can operate in all climates and permit the hot water tank to be located anywhere in the house.

There are still no commercially available solar electric power units which can compete with alternative sources of electricity. Although a great deal of research now underway is aimed at the development of solar-thermal and solar-photovoltaic electric power sources, these units are still in the development stage. Solar cells are being used primarily for supplying small amounts of power in remote locations where conventional sources are not available.

In the area of wind power generation, NASA and ERDA have constructed a 100-kilowatt wind power generator at Sandusky, Ohio. This generator is now being evaluated and it is hoped that generators of this type will be proven economically feasible for certain locations of the country where the average wind velocities are high. Also, wind-powered electric generators are now being manufactured within the U.S. and are generally available at lower costs than foreign units. Only two years ago the U.S. manufactured no wind generators with power outputs of over a kilowatt; now units are available covering the range from several hundred watts (peak) to 15 kilowatts (peak).

At the present time solar energy is more attractive than ever as an alternative energy resource which can provide heating, cooling and hot water at economically competitive costs and without environmental degradation. In order to reduce our nation's dependence on foreign energy resources while maintaining the vitality of our economy, it is imperative that we continue to develop all of our energy options and utilize those which are most appropriate for each particular application. The decision as to what type of energy source should be utilized must, in each case, be made on the basis of economic, environmental and safety considerations. Because of the desirable environmental and safety aspects of solar energy, it should be utilized instead of nuclear or fossil alternatives even when its costs are slightly higher.

Chapter 2

AVAILABILITY OF SOLAR ENERGY

To evaluate the economics and performance of systems for the utilization of solar energy in a particular location, a knowledge of the available solar radiation at that place is essential. Thus, the utilization of solar energy, as with any other natural resource, requires detailed information on availability.

For approximate calculations, average values of energy availability are often used. Solar energy arrives at the surface of the United States at an average rate of about 1500 BTU/ft^2/day (about 42 $\times$ 10^9 BTU/mi^2/day). Over the period of a year a square mile receives about 15 $\times$ 10^{12} BTU. In 1970 the total energy consumed by the U.S. for all purposes was about 6.5 $\times$ 10^{16} BTU. Thus 4300 sq mi of continental U.S. land receives on the average in one year the equivalent of all the U.S. energy needs.[2] At 10% conversion efficiency, 43,000 square miles (about 1.5% of the land area of the 48 contiguous states) could produce the amount of energy the United States consumed in 1970. Solar energy availability has been described as a double periodic function with a 24-hour and a 365-day period length, superimposed with a fluctuating screening function (cloud cover).[17] The maximum amplitude of this function is approximately 300 BTU/ft^2, and for the continental U.S.A. (excluding Alaska) it integrates to an average influx of approximately 570,000 BTU/ft^2/year. Thus, the yearly 24-hour average solar energy received in the United States is about 65 BTU/ft^2 hr.

In order to design a solar energy system, one must calculate the amount of solar energy that will fall on the collector each month. A simple way of doing this is to use the average daily solar radia-

6

tion on a horizontal surface for the closest city listed in Table 1. These values, in BTU/ft²/day, are long-term averages of actual solar radiation based on many years of data, and include the effects of cloud cover. Since Table 1 lists solar intensity data for horizontal surfaces only, a correction must be made for solar collectors that are tilted to the horizontal; the best tilt for collectors used for heating is about 10 degrees greater than the latitude angle, facing southward. Table 2 lists the daily solar radiation incident on a southward-facing surface on a *clear day* as a function of latitude, month and slope. Some interpolation may be required to obtain values for a specific latitude and slope.

Noting the differences between Tables 1 and 2 (Table 1 gives long-term average data including the effects of clouds, Table 2 is for single *clear* days only), these two tables can be used together to estimate the monthly solar radiation available on southward-facing collectors anywhere in the United States. Using Table 2 for the latitude and slope, one calculates for each month the ratio of clear-day solar radiation on the collector to the clear-day solar radiation on a horizontal surface. This ratio is then used to correct the long-term average data (Table 1) for the tilted surface. For example, consider a collector in Columbus, Ohio, with a slope of 50 degrees. The solar radiation on a horizontal surface (Table 1) in January is 486 BTU/ft²/day and the latitude (L) is 40 degrees. Since the slope is 10 degrees greater, using the L + 10 column in Table 2, we find the clear day solar radiation on the collector to be 1906 BTU/ft²/day, and on a horizontal surface it is 948 BTU/ft²/day. Thus we can correct the long-term average horizontal surface value (486) for the tilt by multiplying it by the ratio of 1906 to 948. The solar radiation falling on the tilted surface becomes 486 × (1906/948), or 977 BTU/ft²/day. Thus, it is seen that the solar radiation falling on the tilted surface in January is twice as great as that falling on a horizonal surface.

For more precise calculations, several types of solar radiation data are available including direct radiation at normal incidence, direct plus diffuse radiation at normal incidence, direct radiation on a horizontal surface, direct plus diffuse radiation on a horizontal surface, and each of these on tilted and vertical surfaces.

Table 1. Average Daily Solar Radiation (BTU/ft²/day) on a Horizontal Surface (from Liu and Jordan)

Location	Lat.	Jan	Feb	Mar	Apr	May	Jun	Jul	Aug	Sep	Oct	Nov	Dec
Albuquerque, NM	35°03'	1151	1454	1925	2344	2561	2758	2561	2388	2120	1640	1274	1051
Annette Is., AK	55°02'	236	428	883	1357	1635	1639	1632	1269	962	455	220	152
Apalachicola, FL	29°45'	1107	1378	1654	2041	2269	2196	1979	1913	1703	1545	1243	982
Astoria, OR	46°12'	338	607	1009	1402	1839	1754	2008	1721	1323	780	414	295
Atlanta, GA	33°39'	848	1080	1427	1807	2018	2103	2003	1898	1519	1291	998	752
Barrow, AK	71°20'	13	143	713	1492	1883	2055	1602	954	428	152	23	—
Bismarck, ND	46°47'	587	934	1328	1668	2056	2174	2306	1929	1441	1018	600	464
Blue Hill, MA	42°13'	555	797	1144	1438	1776	1944	1882	1622	1314	941	592	482
Boise, ID	43°34'	519	885	1280	1814	2189	2377	2500	2149	1718	1128	679	457
Boston, MA	42°22'	506	738	1067	1355	1769	1864	1861	1570	1268	897	636	443
Brownsville, TX	25°55'	1106	1263	1506	1714	2092	2289	2345	2124	1775	1537	1105	982
Caribou, ME	46°52'	497	862	1360	1496	1780	1780	1898	1676	1255	793	416	399
Charleston, SC	32°54'	946	1153	1352	1919	2063	2113	1649	1934	1557	1332	1074	952
Cleveland, OH	41°24'	467	682	1207	1444	1928	2103	2094	1841	1410	997	527	427
Columbia, MO	38°58'	651	941	1316	1631	1200	2129	2149	1953	1690	1203	840	590
Columbus, OH	40°00'	486	747	1113	1481	1839	(2111)	2041	1573	1189	920	479	430
Davis, CA	38°33'	599	945	1504	1959	2369	2619	2566	2288	1857	1289	796	551
Dodge City, KS	37°46'	953	1186	1566	1976	2127	2460	2401	2211	1842	1421	1065	874
E. Lansing, MI	42°44'	426	739	1086	1250	1733	1914	1885	1628	1303	892	473	380
East Wareham, MA	41°46'	504	762	1132	1393	1705	1958	1874	1607	1364	997	636	521
Edmonton, AB	53°35'	332	652	1165	1542	1900	1914	1965	1528	1113	704	414	245
El Paso, TX	31°48'	1248	1613	2049	2447	2673	2731	2391	2351	2078	1705	1325	1052
Ely, NV	39°17'	872	1255	1750	2103	2322	2649	2417	2308	1935	1473	1079	815
Fairbanks, AK	64°49'	66	283	861	1481	1806	1971	1703	1248	700	324	104	20
Fort Worth, TX	32°50'	936	1199	1598	1829	2105	2438	2293	2217	1881	1476	1148	914
Fresno, CA	36°46'	713	1117	1653	2049	2409	2642	2512	2301	1898	1416	907	617
Gainesville, FL	29°39'	1037	1325	1635	1945	1935	1961	1896	1874	1615	1312	1170	920
Glasgow, MT	48°13'	573	966	1438	1741	2127	2262	2415	1985	1531	998	575	428

Table 1. Continued

Location	Lat.	Jan	Feb	Mar	Apr	May	Jun	Jul	Aug	Sep	Oct	Nov	Dec
Grand Junction, CO	39°07'	848	1211	1623	2002	2300	2645	2518	2157	1958	1395	970	793
Grand Lake, CO	40°15'	735	1135	1579	1877	1975	2370	2103	1709	1716	1212	776	661
Great Falls, MT	47°29'	524	869	1370	1621	1971	2179	2383	1986	1537	985	575	421
Greensboro, NC	36°05'	744	1032	1323	1755	1989	2111	2034	1810	1517	1203	908	691
Griffin, GA	33°15'	890	1136	1451	1924	2163	2176	2065	1961	1606	1352	1073	782
Hatteras, NC	35°13'	892	1184	1590	2128	2376	2438	2334	2086	1758	1338	1054	798
Indianapolis, IN	39°44'	526	797	1184	1481	1828	2042	2040	1832	1513	1094	662	491
Inyokern, CA	35°39'	1149	1554	2137	2595	2925	3109	2909	2759	2409	1819	1370	1094
Ithaca, NY	42°27'	434	755	1075	1323	1779	2026	2031	1737	1320	918	466	371
Lake Charles, LA	30°13'	899	1146	1487	1802	2080	2213	1969	1910	1678	1506	1122	876
Lander, WY	42°48'	786	1146	1638	1989	2114	2492	2438	2121	1713	1302	837	695
Las Vegas, NV	36°05'	1036	1438	1927	2323	2630	2799	2524	2342	2062	1603	1190	964
Lemont, IL	41°40'	590	879	1256	1482	1866	2042	1991	1837	1469	1016	639	531
Lexington, KY	38°02'	–	–	–	1835	2171	–	2247	2065	1776	1316	–	682
Lincoln, NE	40°51'	713	956	1300	1588	1856	2041	2011	1903	1544	1216	773	643
Little Rock, AR	34°44'	704	974	1336	1669	1960	2092	2081	1939	1641	1283	914	701
Los Angeles, CA	33°56'	931	1284	1730	1948	2197	2272	2414	2155	1898	1373	1082	901
Los Angeles, CA	34°03'	912	1224	1641	1867	2061	2259	2428	2199	1892	1362	1053	878
Madison, WI	43°08'	565	812	1232	1455	1745	3032	2047	1740	1444	993	556	496
Matanuska, AK	61°30'	119	345	–	1328	1628	1728	1527	1169	737	374	143	56
Medford, OR	42°23'	435	804	1260	1807	2216	2441	2607	2262	1672	1044	559	347
Miami, FL	25°47'	1292	1555	1829	2021	2069	1992	1993	1891	1647	1437	1321	1183
Midland, TX	31°56'	1066	1346	1785	2036	2301	2318	2302	2193	1922	1471	1244	1023
Nashville, TN	36°07'	590	907	1247	1662	1997	2149	2080	1863	1601	1224	823	614
Newport, RI	41°29'	566	856	1232	1485	1849	2019	1943	1687	1411	1035	656	528
New York, NY	40°46'	540	791	1180	1426	1738	1994	1939	1606	1349	978	598	476
Oak Ridge, TN	36°01'	604	896	1242	1690	1943	2066	1972	1796	1560	1195	796	610
Oklahoma City, OK	35°24'	938	1193	1534	1849	2005	2355	2274	2211	1819	1410	1086	897

Solar Energy Technology and Applications

Table 1. Continued

Location	Lat.	Jan	Feb	Mar	Apr	May	Jun	Jul	Aug	Sep	Oct	Nov	Dec
Ottawa, ON	45°20'	539	852	1251	1607	1857	2085	2045	1752	1327	827	459	409
Phoenix, AZ	33°26'	1127	1515	1967	2388	2710	2782	2451	2300	2131	1689	1290	1041
Portland, ME	43°39'	566	875	1330	1528	1923	2017	2096	1799	1429	1035	592	508
Rapid City, SD	44°09'	688	1033	1504	1807	2028	2194	2236	2020	1628	1179	763	590
Riverside, CA	33°57'	1000	1335	1751	1943	2282	2493	2444	2264	1955	1510	1169	980
Saint Cloud, MN	45°35'	633	977	1383	1598	1859	2003	2088	1828	1369	890	545	463
Salt Lake City, UT	40°46'	622	986	1301	1813	–	–	–	–	1689	1250	–	553
San Antonio, TX	29°32'	1045	1299	1560	1665	2025	815	2364	2185	1845	1487	1104	955
Santa Maria, CA	34°54'	984	1296	1806	2068	2376	2600	2541	2293	1966	1566	1169	944
Sault Ste. Marie, MI	46°28'	489	844	1337	1559	1962	2064	2149	1768	1207	809	392	360
Sayville, NY	40°30'	603	936	1259	1561	1857	2123	2041	1735	1447	1087	679	534
Schenectady, NY	42°50'	488	754	1027	1272	1553	1688	1662	1495	1125	821	436	357
Seattle, WA	47°27'	283	521	992	1507	1882	1910	2111	1689	1212	702	386	240
Seattle, WA	47°36'	252	472	917	1376	1665	1724	1805	1617	1129	638	326	218
Seabrook, NJ	39°30'	592	854	1196	1519	1801	1965	1950	1715	1446	1072	722	523
Spokane, WA	47°40'	446	838	1200	1765	2104	2227	2480	2076	1511	845	486	279
State College, PA	40°48'	502	749	1107	1399	1755	2028	1968	1690	1336	1017	580	444
Stillwater, OK	36°09'	764	1082	1464	1703	1880	2236	2224	2039	1724	1314	992	783
Tampa, FL	27°55'	1224	1461	1772	2016	2228	2147	1992	1845	1688	1493	1328	1120
Toronto, ON	43°41'	451	675	1089	1388	1785	1942	1969	1623	1284	835	458	353
Tucson, AZ	32°07'	1172	1454	–	2435	–	2601	2292	2180	2123	1641	1322	1132
Upton, NY	40°52'	583	873	1280	1610	1892	2159	2045	1790	1473	1103	687	551
Washington, DC	38°51'	632	902	1255	1600	1847	2081	1930	1712	1446	1083	764	594
Winnipeg, MB	49°54'	488	835	1354	1641	1904	1962	2124	1761	1190	768	445	345

These data are taken from Liu, B. Y. H. and Jordan, R. C. "The Long Term Average Performance of Flat Plate Solar Collectors," published in *Solar Energy*.

Table 2. Clear-Sky Daily Solar Radiation vs Slope and Latitude
(BTU/ft²/day)

Date	Deg. Lat.	I_{DN}	Total Solar Irradiation					
			Horiz.	L-10	L	L+10	L+20	Vertical
Jan. 21	24	2766	1622	1984	2174	2300	2360	1766
	32	2458	1288	1839	2008	2118	2166	1779
	40	2182	948	1660	1810	1906	1944	1726
	48	1710	596	1360	1478	1550	1578	1478
	56	1126	282	934	1010	1058	1074	1044
	64	306	45	268	290	302	306	304
Feb. 21	24	3036	1998	2276	2396	2446	2424	1476
	32	2872	1724	2188	2300	2345	2322	1644
	40	2640	1414	2060	2162	2202	2176	1730
	48	2330	1080	1880	1972	2024	1978	1720
	56	1986	740	1640	1716	1792	1716	1598
	64	1432	400	1230	1286	1302	1282	1252
Mar. 21	24	3078	2270	2428	2456	2412	2298	1022
	32	3012	2084	2378	2403	2358	2246	1276
	40	2916	1852	2308	2330	2284	2174	1484
	48	2780	1578	2208	2228	2182	2074	1632
	56	2586	1268	2066	2084	2040	1938	1700
	64	2296	932	1856	1870	1830	1736	1656
Apr. 21	24	3036	2454	2458	2374	2228	2016	488
	32	3076	2390	2444	2356	2206	1994	764
	40	3092	2274	2412	2320	2168	1956	1022
	48	3076	2106	2358	2266	2114	1902	1262
	56	3024	1892	2282	2186	2038	1830	1450
	64	2982	1644	2776	2082	1936	1736	1594
May 21	24	3032	2556	2447	2286	2072	1800	246
	32	3112	2582	2454	2284	2064	1788	469
	40	3160	2552	2442	2264	2040	1760	724
	48	3254	2482	2418	2234	2010	1728	982
	56	3340	2374	2374	2188	1962	1682	1218
	64	3470	2236	2312	2124	1898	1624	1436
June 21	24	2994	2574	2422	2230	1992	1700	204
	32	3084	2634	2436	2234	1990	1690	370
	40	3180	2648	2434	2224	1974	1670	610
	48	3312	2626	2420	2204	1950	1644	874
	56	3438	2562	2388	2166	1910	1606	1120
	64	3650	2488	2342	2118	1862	1558	1356

Table 2. Continued

Date	Deg. Lat.	I_{DN}	Total Solar Irradiation					
			Horiz.	L-10	L	L+10	L+20	Vertical
July 21	24	2932	2526	2412	2250	2036	1766	246
	32	3012	2558	2442	2250	2030	1754	458
	40	3062	2534	2409	2230	2006	1728	702
	48	3158	2474	2386	2200	1974	1694	956
	56	3240	2372	2342	2152	1926	1646	1186
	64	3372	2248	2280	2090	1864	1588	1400
Aug. 21	24	2864	2408	2402	2316	2168	1958	470
	32	2902	2352	2388	2296	2144	1934	736
	40	2916	2244	2354	2258	2104	1894	978
	48	2898	2086	2300	2200	2046	1836	1208
	56	2850	1883	2218	2118	1966	1760	1392
	64	2808	1646	2108	1008	1860	1662	1522
Sept. 21	24	2878	2194	2432	2366	2322	2212	992
	32	2808	2014	2288	2308	2264	2154	1226
	40	2708	1788	2210	2228	2182	2074	1416
	48	2568	1522	2102	2118	2070	1966	1546
	56	2368	1220	1950	1962	1918	1820	1594
	64	2074	892	1726	1736	1696	1608	1532
Oct. 21	24	2868	1928	2198	2314	2364	2346	1442
	32	2696	1654	2100	2208	2252	2232	1588
	40	2454	1348	1962	2060	2098	2074	1654
	48	2154	1022	1774	1860	1890	1866	1626
	56	1804	688	1516	1586	1612	1588	1480
	64	1238	358	1088	1136	1152	1134	1106
Nov. 21	24	2706	1610	1962	2146	2268	2324	1730
	32	2405	1280	1816	1980	2084	2130	1742
	40	2128	942	1636	1778	1870	1908	1686
	48	1668	596	1336	1448	1518	1544	1442
	56	1094	284	914	986	1032	1046	1016
	64	302	46	266	286	298	302	300
Dec. 21	24	2624	1474	1852	2058	2204	2286	1808
	32	2348	1136	1704	1888	2016	2086	1794
	40	1978	782	1480	1634	1740	1796	1646
	48	1444	446	1136	1250	1326	1364	1304
	56	748	157	620	678	716	734	722
	64	24	2	20	22	24	24	24

For each type of measurement, one may wish to know the maximum and minimum values in selected time periods. It is also necessary to decide what averaging should be employed—seasonal, monthly, daily, hourly, or even shorter intervals. For devices employing focusing systems, data on the intensity and direction of direct radiation would be required. For flat plate collectors the total (direct plus diffuse) radiation intensity on a sloping surface is needed for collectors used in that position. Maximum radiation values are needed to determine the capacity of solar energy systems; evaluation of the system performance over an extended period of time requires data on average intensity and appropriate time-intensity distribution parameters. The form of data most generally used is obtained by continuous monitoring stations which record direct and diffuse solar radiation intensity in a form compatible with digital computers. The form of the data most available and most frequently reported is total radiation (direct plus diffuse) on a horizontal surface received each day (such as is given in Table 1) or in some cases each hour. Approximate methods are available for estimating the direct component and distribution parameters from the total radiation data.

Solar radiation (referred to technically as insolation—incoming solar radiation) is measured by several different types of instruments having various characteristics and degrees of accuracy. Thermoelectric pyranometers, introduced by Kimball and Hobbs in 1923, measure the difference in temperature between black and white surfaces in a glass-enclosed chamber due to the differences in solar-radiation absorption. The electric current output from the thermopiles in these units is recorded on a chart, magnetic tape, or other instrument. If well calibrated and maintained, these instruments can provide long-term and short-term values of solar and sky radiation with an accuracy of 3% or less.

Another type of instrument utilizes the differential expansion of a bimetallic element due to solar absorption to move the stylus on a strip chart recorder. Its accuracy is typically within about 10%. Another radiation instrument is the Bellani pyranometer, which provides an indication of total solar radiation by the quantity of a liquid that has distilled from a solar-heated evaporating chamber. The amount of liquid distilled during a given period provides a

measure of the total insolation received during that period. In the
United States and Europe the thermoelectric pyranometer is most
frequently used. The bimetallic type is simpler and cheaper, and
fairly widely used in South America and Asia, as well as in scattered stations elsewhere in the world.[18]

The number of hours of sunshine per day can be measured by
the Campbell-Stokes sunshine recorder which uses a lens to focus
direct sunshine onto a heat-sensitive paper chart. A discolored line
is produced when the solar disc is focused onto the paper. The
length of the discolored line divided by the total length of the chart
corresponding to the time between sunrise and sunset is the percent
possible sunshine for the day.

Regular measurements of sunshine duration and cloudiness are
made at numerous weather stations throughout the world, and
these records usually cover periods of 20 to 60 years or more. The
average daily radiation is a function of sunshine duration at the
particular location.

The United States National Weather Service solar radiation network presently has over 90 measuring sites. A few of these use
Eppley Model II pyranometers which have about twice the accuracy
of instruments typically used at the other sites. Data are stored at
one-minute intervals on magnetic tape (which can later be processed
by computer). The various primary standards for solar radiation
that have been used have been shown to differ from each other by
as much as 6%, so care must be taken in comparing data from different instruments and from different sites. Instruments which are
being used may degrade as much as 20%–30% before being replaced, so measured intensities can be 20%–30% low for this reason. Some sites, however, have good data with an accuracy of
2%–3%. Much of the data is available from the National Weather
Service as hourly data on tape or cards, and a data format manual
is also available.

Hourly or daily data are no longer published in printed form at
the national level, but only in card, tape or microfilm form. Differences between monthly average sunshine may differ about 40%
from year to year and typically 20%–30% from site to site. There
may be large differences between nearby sites due to local weather

differences, and there can be sizable differences from year to year because of changes in atmospheric turbidity.

Efforts are presently underway to relate reflected solar radiation to ground-level incident radiation so that satellite measurements can be made useful for terrestrial solar energy applications. Absolute deviation of measurements of the solar constant versus wavelength is less than 5% using spectral radiometers.

Surface albedo is determined by taking the 15-day minimum value of reflected sunlight measured by the satellite, and once this value is determined, it can be used to evaluate incident surface radiation from satellite measurements. Satellite measurements should provide very useful data over short time scales, but cannot be extrapolated over long time periods because of variations in surface albedo and atmospheric turbidity. Satellite measurements are needed for microscale data (resolution a few miles). Since interpolation between weather stations is not adequate for specific site studies of solar-thermal conversion this data must either come from satellites or from a continuous insolation monitoring station at the site. One problem, however, is that satellites provide data on total radiation, whereas for systems using solar concentrators, direct beam radiation is needed. One can determine this if the cloudiness is measured, and satellites do measure cloudiness. Recent measurements by NASA determined the solar constant to be 429.0 ± 0.5 BTU/hr/ft^2 (1353 ± 1.5 W/m^2) outside the atmosphere.[19]

The flat plate collector incorporates a transparent cover over a black plate with air or water flowing over or through the black plate, and is usually fixed in position. To evaluate its performance, one must know the intensity, angle and spectrum of solar energy as a function of time. Surface reflectivities depend on the incidence angle, and incident radiation must be split into direct and diffuse components. Empirical techniques have been developed for doing this by using a relationship between daily total radiation outside the atmosphere and daily total at ground level. Statistical distribution curves of hourly radiation versus fraction of time radiation received are very similar for different sites of equivalent overall cloudiness. These data and data on the probability of two consecutive days of cloudiness are needed to determine energy storage requirements of proposed solar energy systems.

One typical insolation measuring station which provides the necessary data for the evaluation of experimental solar energy projects consists of the following equipment:

Kimball-Hobbs thermoelectric pyranometer	$ 560
Spectral pyranometer	990
Normal incidence pyrheliometer	880
Equatorial mount	1225
Wind and temperature instrumentation	1100
Total cost (exclusive of recorders)	$4755

The prices given are 1974 prices as quoted by two well-known vendors. The Kimball-Hobbs pyranometer continuously monitors total insolation in the wavelength range from 0.285 to 2.8 microns. The spectral pyranometer is a similar instrument with hemispherical filters which allow only insolation in restricted wave-length ranges to reach the black and white surfaces, so total (direct plus diffuse) insolation in the restricted wavelength range is measured. The pyrheliometer is supported on the electrically driven equatorial mount and measures direct beam solar radiation only. The wind instrument consists of a low threshold cup anemometer and a vane for measuring wind speed and direction over a speed range of 0.4–100 mph. Three thermisters are used for recording the ambient temperature at heights of 3 meters and 10 meters. This equipment is presently used at an insolation station established at the Georgia Institute of Technology, Atlanta, and provides the necessary insolation and weather data for evaluating the performance of flat plate collectors and other devices that collect solar energy.

Chapter 3

SOLAR ENERGY COLLECTORS

The type of device used to collect solar energy depends primarily on the application. Flat plate thermal energy collectors are used for heating water and heating buildings, but can provide temperatures of only about 150°F above ambient. If higher temperatures are desired, the sunlight must be concentrated onto the collecting surface. If electrical power is to be produced, photovoltaic cells can be used to convert sunlight directly into electricity, either with or without concentrators. The decision as to what kind of collector should be used for a specific application is dictated by technical considerations and economics.

Flat Plate Collectors

Figure 1 illustrates the basic components of a flat plate collector. A black plate is covered by one or more transparent cover plates of glass or plastic, and the sides and bottom of the box are insulated. Sunlight is transmitted through the transparent covers and absorbed by the black surface beneath. The covers tend to be opaque to infrared radiation from the plate, and also retard convective heat transfer from the plate. Thus, the black plate heats up and in turn heats a fluid flowing under, through, or over the plate. Water is most commonly used, since the temperatures involved are usually below the boiling point of water. The hot water may be used directly or for space heating in homes and buildings.

Tests of collectors consisting of two glass panes and a flat black metallic absorber which studied the effects of varying the air gap

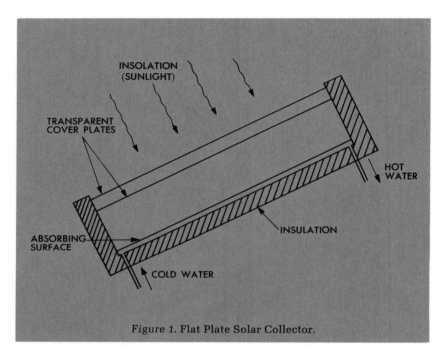

Figure 1. Flat Plate Solar Collector.

between the glass panes and between the glass and collector plate (0.01, 0.02, 0.04 and 0.08 ft) showed that the best performance was with the 0.08-ft spacing, but the performance with the 0.04-ft spacing was almost as good.[20] The performance of these collectors was considerably improved when a selective coating was applied to the collecting surface, instead of flat black paint. Figure 2 illustrates the spectral reflectance of three types of selective coatings.[21] Such coatings strongly absorb incident sunlight, but retard reradiation of infrared heat, and thus allow the collecting surface to reach a higher equilibrium temperature. For a 100°F temperature difference between the outer glass and absorber the collection efficiency increased from 35% to 55% when the selective coating was added, and increased from 10% to 40% when the temperature difference was 150°F.[20]

The collection efficiency of dual glass plate vertical collectors was measured as a function of temperature for three insolation levels. The maximum temperature difference reached was 87°F for

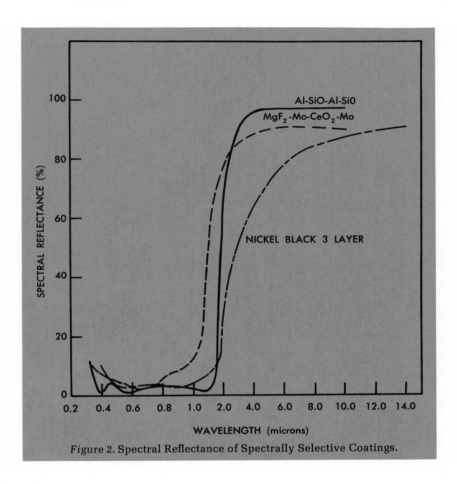

Figure 2. Spectral Reflectance of Spectrally Selective Coatings.

an insolation of 100 BTU/ft²/hr, 153°F at 200 BTU/ft²/hr, and 210°F at 300 BTU/ft²/hr. The collection efficiency was about 50% at half the maximum temperature, and decreased almost linearly to 0 at the maximum temperature.

The efficiency of flat plate collectors can also be improved by anti-reflective coatings on the transparent covers. Figure 3 illustrates the percent of normal incidence sunlight reflected from un-coated and coated glass surfaces.[22] Coated surfaces cost more than uncoated surfaces, of course, and the coating cost increases as the perfomance increases.

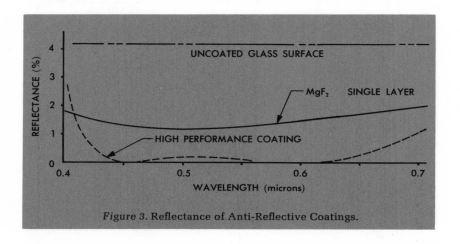

Figure 3. Reflectance of Anti-Reflective Coatings.

Figure 4 illustrates a typical flat plate collector used to provide hot water for space heating and the operation of absorption-type air conditioners. Such collectors are placed on rooftops with a south-ward slope and on south-facing walls. The average daily insolation is reduced about 20% during the winter if the wall faces southeast or southwest instead of south, and is reduced about 60% if the wall faces east or west.

A flat plate collector incorporating solar cells has been de-veloped at the University of Delaware[23] (Figure 5) to supply both electricity and heat for a house. One problem with this type of col-lector is the decrease in photovoltaic conversion efficiency and lifetime with increasing temperature. The 4 × 8 ft collectors are deployed between the roof joists from the inside; the outside is glazed with ¼-in. plexiglas. The heat-transfer fluid for this type of collector is air.

Scientists at NASA's Marshall Space Flight Center have produced a flat plate collector with a new type of selective coating which absorbs over 90% of the incident solar radiation while reemitting only about 6%. A maximum absorber temperature of 450°F was reported without water in the coils.[24]

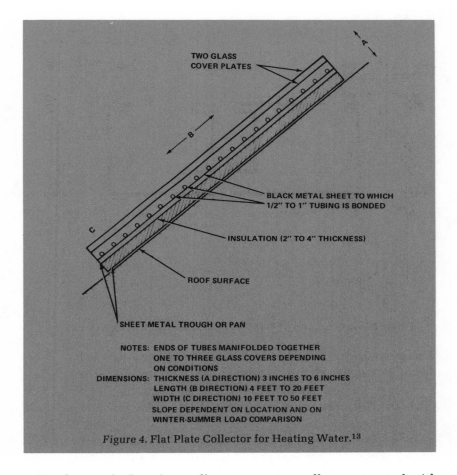

TWO GLASS
COVER PLATES

A

B

BLACK METAL SHEET TO WHICH
1/2" TO 1" TUBING IS BONDED

C

INSULATION (2" TO 4" THICKNESS)

ROOF SURFACE

SHEET METAL TROUGH OR PAN

NOTES: ENDS OF TUBES MANIFOLDED TOGETHER
ONE TO THREE GLASS COVERS DEPENDING
ON CONDITIONS
DIMENSIONS: THICKNESS (A DIRECTION) 3 INCHES TO 6 INCHES
LENGTH (B DIRECTION) 4 FEET TO 20 FEET
WIDTH (C DIRECTION) 10 FEET TO 50 FEET
SLOPE DEPENDENT ON LOCATION AND ON
WINTER-SUMMER LOAD COMPARISON

Figure 4. Flat Plate Collector for Heating Water.[13]

Until recently flat plate collectors were usually constructed with tubes soldered or welded onto a metal plate, which was blackened and covered with one or two cover plates. An exception to this is the Thomason[25,26] collector which uses a corrugated metal collector plate with water flowing on the blackened metal surface between the grooves.

One advance in solar collector technology has been the development of internal-tube collector plates such as the Roll-Bond panel by the Olin Brass Company and the Tube-in-Strip Collector by Revere Copper and Brass. This permits the mass manufacture of absorber plates with integral tubing, and thereby eliminates the

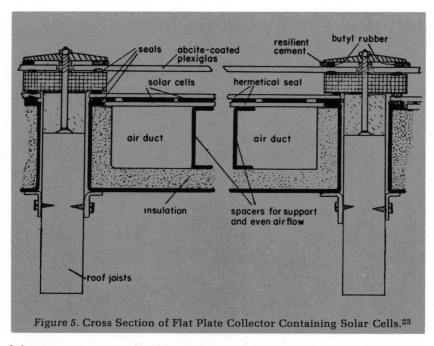

Figure 5. Cross Section of Flat Plate Collector Containing Solar Cells.[23]

laborious process of soldering or welding tubes onto a flat plate. The heat transfer characteristics of the internal tube structure are also superior. Because of these and other desirable features, such absorber plates are being incorporated into many modern collector designs.

One of the first economical flat plate solar collectors using this type of absorber plate was put on the market in 1974 by PPG Industries, Inc. This collector uses two cover plates of 0.32 cm (1/8 in.) Herculite-K tempered glass and aluminum Roll-Bond absorber plate painted with a black enamel which has an absorptivity of about 95%, both in the visible and infrared. It is also available with copper absorber plates and with selective coatings. Fiberglass insulation of 6.35 cm (2.5 in.) thickness is used on the back. The total thickness, less insulation, is 3.5 cm (1-3/8 in.), and the weight is 23.4 kg/m² (4.8 lb/ft²), without fluid, for the aluminum collector. The dimensions of the standard collector are 0.868 m (34-3/16 in.) by 1.935 m (76-3/16 in.), and the cost in February, 1976 was as shown in Table 3. Prices are FOB Ford City, Pennsylvania. Terms are net 30 days. All standard collectors can be shipped within five weeks after receipt of an order and credit approval. Figure 6 illustrates the PPG aluminum collector with insulation and rear cover and Figure 7 shows the copper collector.

Table 3. Prices of PPG Collectors, February 1976

Absorber Plate	Quantity	With Insulation/ & Back Cover	Without Insulation/ & Back Cover
Aluminum	1–7	$214	$175
	8–23	192	158
	24–95	173	142
	96+	160	132
Copper	1–7	$268	$231
	8–23	241	209
	24–95	217	188
	96+	201	173

An attractive integrated roof solar collector for new construction has been developed by Revere Copper and Brass, which has been in the copper roofing business for many years. The attractiveness and good performance of this integrated roof collector, with the collector itself forming a durable leak-tight roof, was demonstrated on the Decade-80 solar house in Tucson. The solar energy system for this house, operating through a complete cooling and heating season, provided 100% of the heating, hot water, and pool heating, and 78% of the cooling of the home. Figure 8 illustrates the construction of the integrated roof on the Decade-80 house, and Figure 9 shows a different tubing arrangement suitable for systems which require a collector that will drain. Revere also sells a modular collector, complete with insulation, for about $9.50 per square foot.

Readers interested in finding out more about the various collector systems available should contact these and other companies that manufacture solar collectors. The list of manufacturers included in the Appendix and your local yellow pages (under "Solar Energy") should be of help.

Research on Flat Plate Solar Collectors

Work at the NASA Marshall Space Flight Center (MSFC) aimed at developing a practical heating and cooling system for buildings

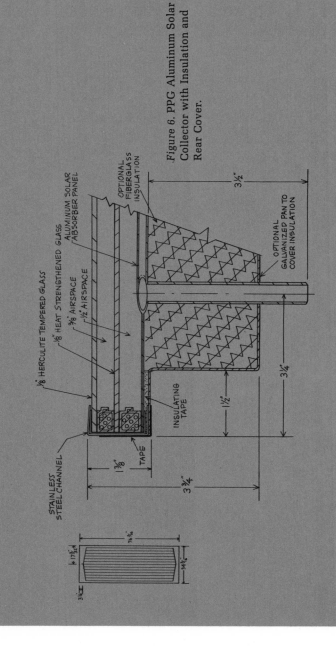

Figure 6. PPG Aluminum Solar Collector with Insulation and Rear Cover.

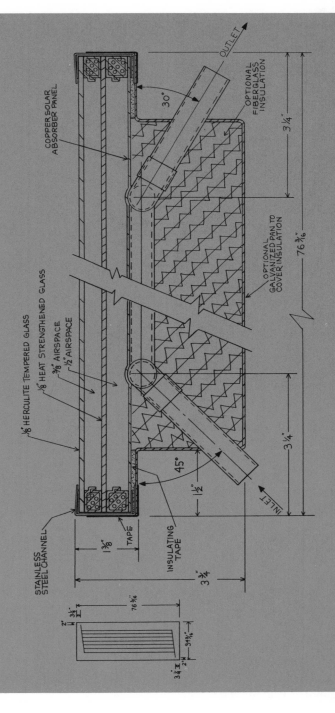

Figure 7. PPG Copper Solar Collector with Insulation and Rear Cover.

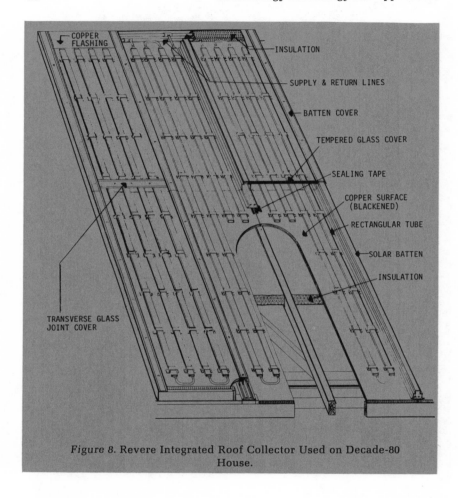

COPPER FLASHING

INSULATION

SUPPLY & RETURN LINES

BATTEN COVER

TEMPERED GLASS COVER

SEALING TAPE

COPPER SURFACE (BLACKENED)

RECTANGULAR TUBE

SOLAR BATTEN

INSULATION

TRANSVERSE GLASS JOINT COVER

Figure 8. Revere Integrated Roof Collector Used on Decade-80 House.

led to the development of an internal tube collector employing reinforced Tedlar films as glazing.[27] The MSFC's plating facilities allowed application of the electroplated selective coating only on panels of small size. Therefore, a 0.61 m (2 ft) wide by 0.91 m (3 ft) long panel size was adopted for the MSFC demonstration system. After selecting the Roll-Bond material and the 2 by 3 ft size of the panels, the flow passage geometry had to be defined. There were three basic considerations involved in designing the flow passages and manifolding for the panels:

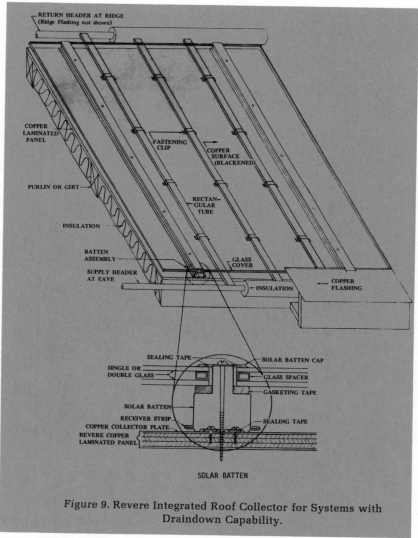

Figure 9. Revere Integrated Roof Collector for Systems with Draindown Capability.

1. Uniform flow distribution throughout the panel is required for good thermal and fluid flow performance.
2. A small overall pressure drop through the panel is required to minimize pump power for fluid circulation.
3. Passage spacing and sizing must be selected for high film efficiency and high film coefficient to achieve good thermal performance.

Parametric fluid flow analyses and thermal analyses were conducted to define a design that would have favorable characteristics with respect to each of the three considerations cited above. For design purposes, a flow rate of 3.46 l/min (0.80 gpm) was assumed to flow through each panel; this was calculated based upon locating seven small panels in series to yield about 3.9 m² (42 ft²) of collector. Under excellent operating conditions this would permit about an 11°C (20°F) rise in water temperature from inlet to outlet. This was taken to be a reasonable value for the maximum temperature rise.

The parametric analyses resulted in a design which would be excellent from fluid distribution, pressure drop and heat transfer viewpoints. This design utilized 16 identical flow passages arranged in parallel. The passages are 0.95 cm (0.375 in.) wide and are spaced on 3.8 cm (1.5 in.) centers. The manifold is a simple triangular passage designed to feed each of the 16 passages with an equal flow of water.

The manifold passage is twice the height of the 16 separate flow passages; i.e., the manifold outside height is 0.64 cm (0.250 in.) rather than the 0.32 cm (0.125 in.) used for the individual flow passages. This extra flow area minimizes pressure variations in the manifold.

The backside of the collector is insulated to minimize heat losses. A transient thermal analysis of this backside insulation was conducted to determine its heat transfer characteristics. Fiberglass insulation having a 15.24 cm (6 in.) thickness was assumed to be subject to a 111°C (200°F) step jump in Roll-Bond plate temperature. The steady-state heat flux of about 24W/m² (8 BTU/ft²/hr) is reached in a few hours. This value was considered acceptable and 15.24 cm (6 in.) of fiberglass insulation was incorporated into the collector design. Parametric studies were conducted to determine the effect of selective coating properties on collector performance under typical operating conditions. The great benefits of high solar absorptance and low infrared emittance were clearly demonstrated. The large benefits offered by the double-glazing were also shown, especially for higher emittance values and lower solar absorptance values. The collectors on the house as well as on the single collector test were inclined at 45° to the horizontal. Long-duration (all-day) tests were run in January and February of 1974 under sunny conditions for performance verification. It had been found earlier that the 0.61 m (2 ft) by 6.4 m (21 ft) test setup was difficult to evaluate in regard to efficiency under transient conditions; i.e., intermittent cloud cover. The predicted efficiencies for various heat fluxes, ambient temperatures and collector average (T_{inlet}-T_{outlet}/2) temperatures, and measured efficiencies are fairly good for the higher solar heat fluxes; however, at the lower heat fluxes it appears that the performance predictions are too low. Since the lower flux data points were obtained as the sun was going down in mid-afternoon, the data are of a transient nature and thermal lag of the collectors may account for the high efficiencies late in the day. Additional testing is required to verify the predictions at the solar fluxes below 946 W/m² (300 BTU/ft²/hr).

The following conclusions have been drawn from the test data on this test collector:

1. Collector efficiency is predicted well by analysis, and the efficiency is high enough to collect the required energy to operate during both winter and summer.
2. The collector can operate at 93°C (200°F) inlet and at 110°C (230°F) outlet at 3.8 l/min (1 gpm) coolant flow (air conditioner temperature requirements).
3. Performance was not degraded measurably in the 2 months that the collector was exposed to winter weather conditions.
4. Based on temperature data, coolant flow through the flow passage of the collector panels is uniform and edge heat leak is not excessive. The temperature increases were not uniform between all panels. This is considered to be a result of measurement inaccuracies.

The Los Alamos Scientific Laboratory developed a collector that uses an absorber plate similar to Roll-Bond, but produced by a welding process.[28] The collector is weather-tight and insulated so that the collector doubles as the roof. It is easily installed and maintained by building craftsmen. Two mild steel plates are welded and pressure-expanded to form a heat-transfer panel. Extensions of the plates provide a structural C-beam support on the bottom and a glazing trough on the top. Glass is added on the top and insulation on the bottom to complete the unit. The collectors are 2 feet wide by 10 feet long, or longer. Factory-assembled cost predictions are $27.00 to $32.30 per square meter ($2.50 to $3.00 per square foot), based on large-scale mass production. A flat-expanded metal surface is made by resistance-welding two thin 0.76 mm (0.03 in. nominal thickness) mild steel plates together with seam welds around the periphery and an array of spot welds over the central portion. Pipe fittings are welded over fluid inlet and exit holes at both ends of one side. The plates are then pressure-expanded to form a flow network of parallel flat flow passages from one end to the other. The resulting surface has a quilted effect and is surface-coated with a selective and protective coating. Use of mild steel necessitates that the corrosion problem be solved with respect to the internal fluids as well as the external coatings. This problem has many solutions which may produce a mild steel collector of long life and high reliability.

The PPG, Revere, NASA and LASL collectors which have been described are examples of a variety of collectors that have been introduced.

Selective Coatings

Spectrally selective coatings applied to solar collector absorber plates enhance the absorption of sunlight while reducing the emission of infrared radiation, thus causing a higher equilibrium temperature to be reached. At a given operating temperature, the collector efficiency is increased because less heat is lost by reradiation.

Necessary properties of a practical solar selective coating are ease and availability of application, low cost, and long-term durability under solar radiation. One method of producing a solar selective coating has been to coat the collector surface with black copper oxide by oxidation of a copper surface on the collector or by thermal decomposition of copper nitrate on the collector surface. Another solar selective coating was discovered by Tabor who found that electrodeposited black nickel had solar selective properties. Both of these coatings, in correct application, have high absorptance in the visible and low emissivity in the infrared.

Work at the NASA-Lewis Research Center[29] has determined

that a durable (commercially widely available) decorative electroplated finish of black chrome has desirable solar selective properties of high absorptance in the visible and low emissivity in the
infrared. This black chrome electroplating solution can be prepared
independently or is available as a proprietary mixture from Harshaw Chemical Co. To investigate the adaptability of electroplated
black chrome to solar collectors, samples varying from 10.16 ×
15.24 cm (4 × 6 in.) to collector tube sheets 0.61 × 1.22 m (2 × 4
ft) were prepared at NASA/Lewis. Visible and infrared spectral
reflectance was measured to determine the quality of the solar selective coating. The 0.61 × 1.22 m (2 × 4 ft) panels were plated by
commercial electroplaters at a cost of about \$7.50/m² (0.70/ft²)
using the same tanks, chemicals, and procedures used for decorative finish black chrome such as table legs and metal chair trim.
The black chrome was prepared on both steel and aluminum to
determine adaptability to various substrates.

The black chrome deposits used in spectral measurements of
solar selective properites were electroplated on 10.16 × 15.24 cm
(4 × 6 in.) test panels by the Harshaw Chemical Co. These deposits
were standard bright, decorative black chrome, a highly specular
black mirror. The panels were 0.89 cm (0.035 in.) cold rolled steel
buffed to less than 0.0127 microns RMS finish. The spectral reflectance from 0.35 to 2.1 microns of the black chrome was measured with a spectrophotometer with a spherical diffuse reflector
attachment. A MgO surface, prepared at the NASA/Lewis Research
Center, was used as a standard. All measurements reported, 0.35 to 2.1
microns were total diffuse reflectance.

The spectral reflectance from 3.0 to 18.0 microns was measured
with a spectrophotometer which uses a spherical diffuse reflectance
attachment. Evaporated gold film was used as a standard. All measurements reported are total diffuse reflectance. The application of
black chrome was determined to be equally feasible on aluminum
base or on steel base. It can also be applied to copper.

The general appearance of black chrome is so indistinguishable
from black nickel that a 0.61 × 1.22 m (2 × 4 foot) solar collector
panel coated with black chrome by the NASA and a second solar
collector of equal size coated with black nickel from an outside
source were indistinguishable by visual observation by any of a

number of people when the panels were placed side by side. Indeed, since the mechanical design of the black chrome coated panel and the black nickel coated panel were identical, the panels could not be separated by visual appearance, and it was only after secondary markings were checked that the two panels could be correctly identified.

The reflectance versus wavelength increases more rapidly in the infrared for black nickel than for the measured specimen of black chrome. However, the reflectance of the black chrome and the first sample of black nickel are significantly higher than that of the second sample of black nickel. This tends to indicate that variables in coating formation process are more significant than inherent difference between black chrome and black nickel.

The values of absorptivity, α, and emissivity, ϵ, and the ratio, α/ϵ, are presented in Table 4. Values from 3 to 15 microns are integrated values over blackbody thermal spectrum. The α/ϵ values determined for the black chrome, two samples of black nickel, and black paint are 9.8, 13.3, 8.0, and 1.01, respectively. Commercial production samples of black chrome nonoptimized for solar selectivity have solar absorptance and infrared emissivity characteristics within the range of properties of black nickel produced for solar use.

Table 4. Visible Absorptance and Infrared Emissivity of Solar Selective Coatings

Coating Sample Number	Absorptivity α (air mass 2)	Emissivity, ε (blackbody integration 3 to 15 microns)	$\alpha^*/\varepsilon^{**}$
Black chrome 1	0.868	0.088	9.8
Black nickel 2	0.877	0.066	13.3
Black nickel 3	0.867	0.109	8.0
Nextel black paint 4	0.967	0.967	1.0

* Based on Solar air-mass-2 spectrum weightings.
** Based on 250°F blackbody spectrum weightings.

Solar Concentrators

Concentrators may be used to produce temperatures in excess of 300°F for efficient electrical power generation, for industrial and agricultural drying, and for other applications where high-temperature heat is needed. Also, concentrators have been used to increase the power output of photovoltaic cells.

One method of concentrating the sun's rays is through a lens. A small magnifying glass will concentrate enough heat to burn wood or paper. Making a large enough lens to concentrate a large amount of incident radiation is difficult, so most solar concentrators employ a reflector system. For high concentration the ideal form of the concentrator, from an optical standpoint, is parabolic; however, to achieve this high concentration the reflector must be steered to remain directed toward the sun, and the heat exchanger must remain located at its focus.[30] For this reason, parabolic concentrators are seldom considered for most solar energy applications where the cost of collecting the solar energy must be kept low. Large solar collectors are subject to large wind loadings, and thus require a sturdy supporting structure.

Since the sun has an apparent diameter with an angle of approximately 0.009 radians, the ideal parabola will concentrate most of the radiation on a hot spot with a diameter equal to the focal length times 0.009 radians. The concentration ratio C produced at the hot spot is defined as the ratio of the solar radiation intensity on the hot spot to the unconcentrated direct sunshine intensity at the concentrator site:

$$C = \frac{q_f}{q_i} = \frac{\text{solar radiation intensity at hot spot}}{\text{unconcentrated direct solar radiation}}$$

For a perfect paraboloid the ratio is found to be a function of the paraboloid rim angle θ and the angular diameter of the sun (a = 0.009 radians).

$$C = \frac{4}{a^2} \sin^2 \theta$$

or:

$$q_f = q_i \frac{4}{a^2} \sin^2 \theta$$

The parabola must be oriented with its axis toward the sun. One method of maintaining this orientation is to rotate the parabolic concentrator tracking the sun's motion across the sky, and move it about two axes to account for seasonal as well as daily motion.

An alternative to steering the concentrator is to use auxiliary mirrors. In this approach a large flat plane mirror is used to track

the sun and reflect its rays into the parabolic concentrator. The surface of the plane mirror must be of greater area than the paraboloid
because of the angles involved, but its structure is less complex
than the paraboloid. Because the reflecting surface efficiency of
any surface is less than 100%, the use of auxiliary mirrors results
in an efficiency loss. The cost factor along with the advantages of
a fixed working surface make auxiliary mirrors common in modern
parabolic solar concentrators.

For a solar concentrator the maximum temperature obviously
cannot exceed the temperature of the sun. Theoretically, for an
ideal paraboloid of 100% reflectivity operating in space, the temperature of a black body at the hot spot would reach approximately
10,000°F. However, the atmosphere reduces the incident radiation
of the sun. In the United States the ratio of average total incident
radiation received to that received outside the atmosphere ranges
from over 0.75 to under 0.40. The reflectivity of aluminum ranges
from 0.89 to 0.53, and for mirror glass this reflectivity ranges from
0.72 to 0.96. These factors along with other irregularities limit the
reported temperatures for existing solar furnaces to about 7000°F.
A small parabolic concentrator is capable of producing the same
temperature as a large concentrator, the difference being only the
amount of heat collected and the size of the hot spot. The hot spot
on most moderate-to-large-size furnaces has a diameter of 1–2 in.
To produce a hot spot of 5-in. diameter requires a reflector with a
diameter equal to the height of a 10-story building. Concentrators
using toroidal, flat or spherical components are usually cheaper to
produce than parabolic concentrators.[31]

One simple type of concentrating solar collector (Figure 10) uses a
parabolic cylinder reflector to concentrate sunlight onto a collecting
pipe within a quartz or pyrex envelope. The pipe can be coated with
a selective coating (Figure 2) to retard infrared emission, and the
transparent tube surrounding the pipe can be evacuated to reduce
convective heat losses. The reflector is steered during the day to
keep sunlight focused on the collector. This type of concentrator,
known as the parabolic trough concentrator, cannot produce as high
a temperature as the parabolic reflector, but produces much higher
temperatures than flat plate collectors.

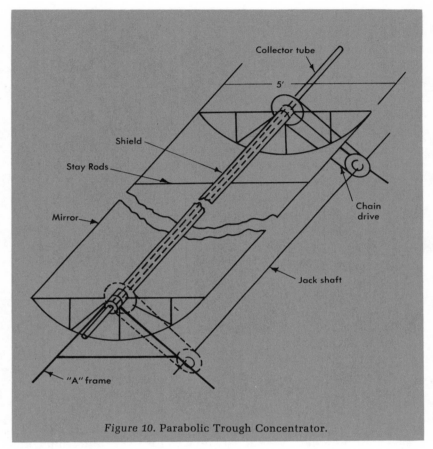

Figure 10. Parabolic Trough Concentrator.

Solar-thermal collectors may be categorized as (1) low-tempera-
ture flat plate collectors with no concentration, (2) medium-tem-
perature concentrating collectors typified by parabolic cylinders,
and (3) high-concentration, high-temperature collectors such as
parabolic concentrators or concentrators composed of many flat
mirrors focused at the same point. The actual temperature obtained
will depend on the optical performance of the reflector, the accuracy of the tracking device, and the absorption efficiency of the
receiver.[32]

Researchers in the Soviet Union have developed a technique for
mass producing inexpensive faceted solar concentrators which
form an approximate parabolic cylinder. They used a jig containing

a number of vacuum socket facet holders, arranged along a convex cylindrical parabolic surface, all connected to a central vacuum system. In making a concentrator, the 26 mirror strips were placed face down on the correctly positioned holders and the vacuum held the mirror facets in the desired position throughout the manufacturing process. The reverse side of the mirrors was then coated with a layer of epoxy resin and covered with glass fabric. The supporting structure, which had the approximate surface shape of the finished concentrator, was placed on the glass fabric and glued to the mirror. After the epoxy had cured, the vacuum was turned off and the finished concentrator removed. Soviet researchers manufactured 80 concentrator sections one meter long by about one meter wide using this technique. These concentrator sections were used to make two power plants. It was only necessary to align the sections, and not the individual facets. These concentrators were cheap to produce, had good optical characteristics, and were quite strong.[33]

Two general approaches have been used to try to reduce the expense and engineering difficulties associated with steering the reflecting surface: (1) use simple automatic steering mechanisms to move many separate reflectors, which require less supporting structure than a single large concentrator, and (2) develop concentrators with fixed mirrors and movable heat collectors.

Concentrator Systems

Concentrators offer several advantages for the heating and cooling of buildings:

1. Higher collection efficiencies result in smaller collectors
2. More compact heat storage
3. Year round collection of high-temperature heat
4. More efficient operation of absorption cooling devices

Also, higher-temperature heat collection makes the generation of electric power possible, with waste heat used for space heating and air conditioning. With concentrators air or a commercial heat transfer fluid will be used.

Instead of steering a single concentrator Gunter[34] proposed a faceted solar concentrator in which the separate flat reflecting facets were rotated by a single mechanism. Each facet is rotated at exactly the same speed to keep the reflected sunlight focused on a fixed heat-collecting element. Another approach is to focus many separate flat mirrors onto a single point. The difficulty with this latter system is that each mirror requires a separate steering mechanism; but if large numbers are used, they may lend themselves to the economics of mass production.

The second approach to reducing the concentrator cost is to fix the reflector and move the heat-collecting element. The problem with this is that the standard reflecting surfaces are only in focus for one of the sun's directions. The parabolic cylinder and parabolic concentrators are only in focus when the sunlight is incident along the axis of the parabola, so the problem with such fixed collectors, as proposed by Steward,[35] is that the focus is severely degraded whenever the incident direction of the sunlight is significantly off axis. This results in a reduced concentration factor and reduced collection temperature.

The faceted fixed mirror concentrator[36] remains in focus for any incident sun angle. It is composed of long, narrow, flat reflecting elements arranged on a concave cylindrical surface. The angles of the reflecting elements are fixed so that the focal distance is twice the radius of the cylindrical surface. The focus is always sharp for parallel light of any incident direction. The point of focus lies on the reference cylindrical surface, so the heat exchanger pipe can be supported on arms that pivot at the center of the reference cylinder. This greatly simplifies the positioning of the heat exchanger.

Chapter 4

HEATING FOR HOUSES AND BUILDINGS

The Committee on Science and Astronautics of the U.S. House of Representatives has concluded that "the most promising area for the application of solar energy within the next 10–15 years, on a scale sufficient to yield measurable relief from the increasing demands upon fossil fuels and other conventional energy sources, is the use of solar energy for space heating, air conditioning, and water heating in buildings."[37] As is seen from Table 5, energy for space heating, air conditioning, and water heating in building services accounts for about 25% of the total energy consumption in the United States, and is presently supplied almost totally by the

Table 5. Energy Consumption in the United States by End Use, 1960–68

(Trillions of BTU and Percent/Year)

Sector and End Use	Consumption		Annual Rate of Growth (%)	Percent of National Total	
	1960	1968		1960	1968
Residential:					
Space heating	4,848	6,675	4.1	11.3	11.0
Water heating	1,159	1,736	5.2	2.7	2.9
Cooking	556	637	1.7	1.3	1.1
Clothes drying	93	208	10.6	.2	.3
Refrigeration	369	692	8.2	.9	1.1
Air conditioning	134	427	15.6	.3	.7
Other	809	1,241	5.5	1.9	2.1
Total	7,968	11,616	4.8	18.6	19.2

Table 5. Continued

Sector and End Use	Consumption 1960	Consumption 1968	Annual Rate of Growth (%)	Percent of National Total 1960	Percent of National Total 1968
Commercial:					
Space heating	3,111	4,182	3.8	7.2	6.9
Water heating	544	653	2.3	1.3	1.1
Cooking	98	139	4.5	.2	.2
Refrigeration	534	670	2.9	1.2	1.1
Air conditioning	576	1,113	8.6	1.3	1.8
Feedstock	734	984	3.7	1.7	1.6
Other	145	1,025	28.0	.3	1.7
Total	5,742	8,766	5.4	13.2	14.4
Industrial:					
Process Steam	7,646	10,132	3.6	17.8	16.7
Electric drive	3,170	4,794	5.3	7.4	7.9
Electrolytic processes	486	705	4.8	1.1	1.2
Direct heat	5,550	6,929	2.8	12.9	11.5
Feedstock	1,370	2,202	6.1	3.2	3.6
Other	118	198	6.7	.3	.3
Total	18,340	24,960	3.9	42.7	41.2
Transportation:					
Fuel	10,873	15,038	4.1	25.2	24.9
Raw materials	141	146	.4	.3	.3
Total	11,014	15,184	4.1	25.5	25.2
National Total	43,064	60,526	4.3	100.0	100.0

Note: Electric Utility consumption has been allocated to each end use.
Source: Patterns of Energy Consumption in the United States.[14]

combustion of high-quality fossil fuels. The sources which supply this energy are depicted by Figure 11. Space heating accounts for more than half of the total residential energy consumption. Space heating alone for homes and businesses accounts for 18% of all energy consumption in the United States. In the South, where solar energy is most available, practically all residential energy comes from gas or electricity, and even in the South about half this energy is used for space heating. Space heating and water heating account for over two-thirds of all residential energy consumption.

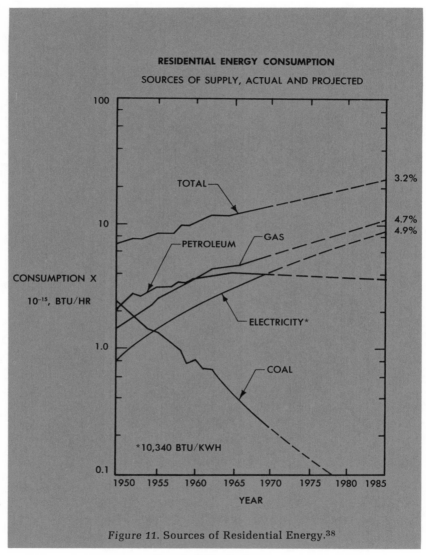

Figure 11. Sources of Residential Energy.[38]

Load Calculations

In order to design a solar system for a particular home or building, the amount of heating or cooling required by the building must be calculated. This calculation is quite different from what is usually done for conventional HVAC systems. The purpose of a load calculation for a conventional building is to determine only the maximum heating or cooling load (on the coldest or hottest day of the year) and then the furnace or air conditioning unit is sized to meet this load. For a solar system, the total energy required for heating and cooling, on a month-by-month or day-by-day basis, is required.

Experience has shown that if the indoor temperature of a structure is maintained at 70°F, there is no space heating required if the outside ambient temperature is at least 65°F. Heating is required if the ambient temperature is less than 65°F, and the amount of heat input is proportional to the difference between the outside temperature and 65°F. In calculating long-term heating requirements, the use of tabulated degree-day data is convenient. If the average outside temperature on a particular day is 30°F, then the average difference between 30°F and 65°F is 35°F, and since this average temperature difference occurred for one day, then the load for that day is calculated based on 35 degree-days. If the following day the average ambient temperature was 40°F, then we have 25 heating degree-days, for a total of 60 heating degree-days for the 2-day period. An example for an entire week is given in Table 6.

Table 6. Calculation of Heating Degrees Days for a Week

	Average Outside Temperature (°F)	Degree-Days (°F day)
Sunday	30	35
Monday	40	25
Tuesday	55	10
Wednesday	65	0
Thursday	70	0
Friday	45	20
Saturday	25	40
Total for Week		130

Note that whenever the average outside temperature is 65° or higher, the heating degree-days are zero. Monthly total degree-day data have been compiled for cities across the United States by the Environmental Data Service of the U.S. Department of Commerce National Oceanic and Atmospheric Administration. Some of these data are given in Table 7. The monthly heating requirement is then

Table 7. Normal Total Heating Degree-Days (Base 65°)

State/City	Jan	Feb	Mar	Apr	May	Jun	Jul	Aug	Sep	Oct	Nov	Dec	Year
Alabama													
Birmingham	592	462	363	108	9	0	0	0	6	93	363	555	2551
Huntsville	694	557	434	138	19	0	0	0	12	127	426	663	3070
Mobile	415	300	211	42	0	0	0	0	0	22	213	357	1560
Montgomery	543	417	316	90	0	0	0	0	0	68	330	527	2291
Alaska													
Anchorage	1631	1316	1293	879	592	315	245	291	516	930	1284	1572	10864
Annette	949	837	843	648	490	321	242	208	327	567	738	899	7069
Barrow	2517	2332	2468	1944	1445	957	803	840	1035	1500	1971	2362	20174
Barter Island	2536	2369	2477	1923	1373	924	735	775	987	1482	1944	2337	19862
Bethel	1903	1590	1655	1173	806	402	319	394	612	1042	1434	1866	13196
Cold Bay	1153	1036	1122	951	791	591	474	425	525	772	918	1122	9880
Cordova	1299	1086	1113	864	660	444	366	391	522	781	1017	1221	9764
Fairbanks	2359	1901	1739	1068	555	222	171	332	642	1203	1833	2254	14279
Juneau	1237	1070	1073	810	601	381	301	338	483	725	921	1135	9075
King Salmon	1600	1333	1411	966	673	408	313	322	513	908	1290	1606	11343
Kotzebue	2192	1932	2080	1554	1057	636	381	446	723	1249	1728	2127	16105
McGrath	2294	1817	1758	1122	648	258	208	338	633	1184	1791	2232	14283
Nome	1879	1666	1770	1314	930	573	481	496	693	1094	1455	1820	14171
Saint Paul	1228	1168	1265	1098	936	726	605	539	612	862	963	1197	11199
Shemya	1045	958	1011	885	837	696	577	475	501	784	876	1042	9687
Yakutat	1169	1019	1042	840	632	435	338	347	474	716	936	1144	9092

Arizona													
Flagstaff	1169	991	911	651	437	180	46	68	201	558	867	1073	7152
Phoenix	474	328	217	75	0	0	0	0	0	22	234	415	1765
Prescott	865	711	605	360	158	15	0	0	27	245	579	797	4362
Tucson	471	344	242	75	6	0	0	0	0	25	231	406	1800
Winslow	1054	770	601	291	96	0	0	0	6	245	711	1008	4782
Yuma	363	228	130	29	0	0	0	0	0	0	148	319	1217
Arkansas													
Fort Smith	781	596	456	144	22	0	0	0	12	127	450	704	3292
Little Rock	756	577	434	126	9	0	0	0	9	127	465	716	3219
Texarkana	626	468	350	105	0	0	0	0	0	78	345	561	2533
California													
Bakersfield	546	364	267	105	19	0	0	0	0	37	282	502	2122
Bishop	874	666	539	306	143	36	0	0	42	248	576	797	4227
Blue Canyon	865	781	791	582	397	195	34	50	120	347	579	766	5507
Burbank	366	277	239	138	81	18	0	0	6	43	177	301	1646
Eureka	546	470	505	438	372	285	270	257	258	329	414	499	4643
Fresno	586	406	319	150	56	0	0	0	0	78	339	558	2492
Long Beach	375	297	267	168	90	18	0	0	12	40	156	288	1711
Los Angeles	372	302	288	219	158	81	28	22	42	78	180	291	2061
California (cont)													
Mt. Shasta	983	784	738	525	347	159	25	34	123	406	696	902	5722
Oakland	527	400	255	355	180	90	53	50	45	127	309	481	2870
Point Arguello	474	392	403	339	298	243	186	202	162	205	291	400	3595
Red Bluff	605	428	341	168	47	0	0	0	0	53	318	555	2515
Sacramento	614	442	360	216	102	6	0	0	12	81	363	577	2773
Sandberg	778	661	620	426	264	57	0	0	30	202	480	691	4209
San Diego	313	249	202	123	84	36	6	0	15	37	123	251	1439
San Francisco	508	395	363	279	214	126	81	78	60	143	306	462	3015
Santa Catalina	353	308	326	249	192	105	16	0	9	50	165	279	2052
Santa Maria	459	370	363	282	233	165	99	93	96	146	270	391	2967

State/City	Jan	Feb	Mar	Apr	May	Jun	Jul	Aug	Sep	Oct	Nov	Dec	Year
Colorado													
Alamosa	1476	1162	1020	696	440	168	65	99	279	639	1065	1420	8529
Colorado Springs	1128	938	893	582	319	84	9	25	132	456	825	1032	6423
Denver	1132	938	887	558	288	66	6	9	117	428	819	1035	6283
Grand Junction	1209	907	729	387	146	21	0	0	30	313	786	1113	5641
Pueblo	1085	871	772	429	174	15	0	0	54	326	750	986	5462
Connecticut													
Bridgeport	1079	966	853	510	208	27	0	0	66	307	615	986	5617
Hartford	1209	1061	899	495	177	24	0	6	99	372	711	1119	6172
New Haven	1097	991	871	543	245	45	0	12	87	347	648	1011	5897
Delaware													
Wilmington	980	874	735	387	112	6	0	0	51	270	588	927	4930
Florida													
Apalachicola	347	260	180	33	0	0	0	0	0	16	153	319	1308
Daytona Beach	248	190	140	15	0	0	0	0	0	0	75	211	879
Fort Myers	146	101	62	0	0	0	0	0	0	0	24	109	442
Jacksonville	332	246	174	21	0	0	0	0	0	12	144	310	1239
Key West	40	31	9	0	0	0	0	0	0	0	0	28	108
Lakeland	195	146	99	0	0	0	0	0	0	0	57	164	661
Miami Beach	56	36	9	0	0	0	0	0	0	0	0	40	141
Orlando	220	165	105	6	0	0	0	0	0	0	72	198	766
Pensacola	400	277	183	36	0	0	0	0	0	19	195	353	1463
Tallahassee	375	286	202	36	0	0	0	0	0	28	198	360	1485
Tampa	202	148	102	0	0	0	0	0	0	0	60	171	683
West Palm Beach	87	64	31	0	0	0	0	0	0	0	6	65	253

Georgia													
Athens	642	529	431	141	22	0	0	0	12	115	405	632	2929
Atlanta	639	529	437	168	25	0	0	0	18	127	414	626	2983
Augusta	549	445	350	90	0	0	0	0	0	78	333	552	2397
Columbus	552	434	338	96	0	0	0	0	0	87	333	543	2383
Macon	505	403	295	63	0	0	0	0	0	71	297	502	2136
Rome	710	577	468	177	34	0	0	0	24	161	474	701	3326
Savannah	437	353	254	45	0	0	0	0	0	47	246	437	1819
Thomasville	394	305	208	33	0	0	0	0	0	25	198	366	1529
Idaho													
Boise	1113	854	722	438	245	81	0	0	132	415	792	1017	5809
Idaho Falls 46W	1538	1249	1085	651	391	192	16	34	270	623	1056	1370	8475
Idaho Falls 42NW	1600	1291	1107	657	388	192	16	40	282	648	1107	1432	8760
Lewiston	1063	815	694	426	239	90	0	0	123	403	756	933	5542
Pocatello	1324	1058	905	555	319	141	0	0	172	493	900	1166	7033
Illinois													
Cairo	856	680	539	195	47	0	0	0	36	164	513	791	3821
Chicago	1209	1044	890	480	211	48	0	0	81	326	753	1113	6155
Moline	1314	1100	918	450	189	39	0	9	99	335	774	1181	6408
Peoria	1218	1025	849	426	183	33	0	6	87	326	759	1113	6025
Rockford	1333	1137	961	516	236	60	6	9	114	400	837	1221	6830
Springfield	1135	935	769	354	136	18	0	0	72	291	696	1023	5429
Indiana													
Evansville	955	767	620	237	68	0	0	0	66	220	606	896	4435
Fort Wayne	1178	1028	890	471	189	39	0	9	105	378	783	1135	6205
Indianapolis	1113	949	809	432	177	39	0	0	90	316	723	1051	5699
South Bend	1221	1070	933	525	239	60	0	6	111	372	777	1125	6439

State/City	Jan	Feb	Mar	Apr	May	Jun	Jul	Aug	Sep	Oct	Nov	Dec	Year
Iowa													
Burlington	1259	1042	859	426	177	33	0	0	93	322	768	1135	6114
Des Moines	1398	1165	967	489	211	39	0	9	99	363	837	1231	6808
Dubuque	1420	1204	1026	546	260	78	12	31	156	450	906	1287	7376
Sioux City	1435	1198	989	483	214	39	0	9	108	369	867	1240	6951
Waterloo	1460	1221	1023	531	229	54	12	19	138	428	909	1296	7320
Kansas													
Concordia	1163	935	781	372	149	18	0	0	57	276	705	1023	5479
Dodge City	1051	840	719	354	124	9	0	0	33	251	666	939	4986
Goodland	1166	955	884	507	236	42	0	6	81	381	810	1073	6141
Topeka	1122	893	722	330	124	12	0	0	57	270	672	980	5182
Wichita	1023	804	645	270	87	6	0	0	33	229	618	905	4620
Kentucky													
Covington	1035	893	756	390	149	24	0	0	75	291	669	983	5265
Lexington	946	818	685	325	105	0	0	0	54	239	609	902	4683
Louisville	930	818	682	315	105	9	0	0	54	248	609	890	4660
Louisiana													
Alexandria	471	361	260	69	0	0	0	0	0	56	273	431	1921
Baton Rouge	409	294	208	33	0	0	0	0	0	31	216	369	1560
Burrwood	298	218	171	27	0	0	0	0	0	0	96	214	1024
Lake Charles	381	274	195	39	0	0	0	0	0	19	210	341	1459
New Orleans	363	258	192	39	0	0	0	0	0	19	192	322	1385
Shreveport	552	426	304	81	0	0	0	0	0	47	297	477	2184
Maine													
Caribou	1690	1470	1308	858	468	183	78	115	336	682	1044	1535	9767
Portland	1339	1182	1042	675	372	111	12	53	195	508	807	1215	7511

Location													
Maryland													
Baltimore	936	820	679	327	90	0	0	0	48	264	585	905	4654
Frederick	995	876	741	384	127	12	0	0	66	307	624	955	5087
Massachusetts													
Blue Hill Obsy.	1178	1053	936	579	267	69	0	22	108	381	690	1085	6368
Boston	1088	972	846	513	208	36	0	9	60	316	603	983	5634
Nantucket	992	941	896	621	384	129	12	22	93	332	573	896	5891
Pittsfield	1339	1196	1063	660	326	105	25	59	219	524	831	1231	7578
Worcester	1271	1123	998	612	304	78	6	34	147	450	774	1172	6969
Michigan													
Alpena	1404	1299	1218	777	446	156	68	105	273	580	912	1268	8506
Detroit	1181	1058	936	522	220	42	0	0	87	360	738	1088	6232
Escanaba	1445	1296	1203	777	456	159	59	87	243	539	924	1293	8481
Flint	1330	1198	1066	639	319	90	16	40	159	465	843	1212	7377
Grand Rapids	1259	1134	1011	579	279	75	9	28	135	434	804	1147	6894
Lansing	1262	1142	1011	579	273	69	6	22	138	431	813	1163	6909
Marquette	1411	1268	1187	771	468	177	59	81	240	527	936	1268	8393
Muskegon	1209	1100	995	594	310	78	12	28	120	400	762	1088	6696
Sault Ste. Marie	1525	1380	1277	810	477	201	96	105	279	580	951	1367	9048
Minnesota													
Duluth	1745	1518	1355	840	490	198	71	109	330	632	1131	1581	10000
International Falls	1919	1621	1414	828	443	174	71	112	363	701	1236	1724	10606
Minneapolis	1631	1380	1166	621	288	81	22	31	189	505	1014	1454	8382
Rochester	1593	1366	1150	630	301	93	25	34	186	474	1005	1438	8295
Saint Cloud	1702	1445	1221	666	326	105	28	47	225	549	1065	1500	8879
Mississippi													
Jackson	546	414	310	87	0	0	0	0	0	65	315	502	2239
Meridian	543	417	310	81	0	0	0	0	0	81	339	518	2289
Vicksburg	512	384	282	69	0	0	0	0	0	53	279	462	2041

State/City	Jan	Feb	Mar	Apr	May	Jun	Jul	Aug	Sep	Oct	Nov	Dec	Year
Missouri													
Columbia	1076	874	716	324	121	12	0	0	54	251	651	967	5046
Kansas City	1032	818	682	294	109	0	0	0	39	220	612	905	4711
St. Joseph	1172	949	769	348	133	15	0	6	60	285	708	1039	5484
St. Louis	1026	848	704	312	121	15	0	0	60	251	627	936	4900
Springfield	973	781	660	291	105	6	0	0	45	223	600	877	4561
Montana													
Billings	1296	1100	970	570	285	102	6	15	186	487	897	1135	7049
Glasgow	1711	1439	1187	648	335	150	31	47	270	608	1104	1466	8996
Great Falls	1349	1154	1063	642	384	186	28	53	258	543	921	1169	7750
Havre	1584	1364	1181	657	338	162	28	53	306	595	1065	1367	8700
Helena	1438	1170	1042	651	381	195	31	59	294	601	1002	1265	8129
Kalispell	1401	1134	1029	639	397	207	50	99	321	654	1020	1240	8191
Miles City	1504	1252	1057	579	276	99	6	6	174	502	972	1296	7723
Missoula	1420	1120	970	621	391	219	34	74	303	651	1035	1287	8125
Nebraska													
Grand Island	1314	1089	908	462	211	45	0	6	108	381	834	1172	6530
Lincoln	1237	1016	834	402	171	30	0	6	75	301	726	1066	5864
Norfolk	1414	1179	983	498	233	48	9	0	111	397	873	1234	6979
North Platte	1271	1039	930	519	248	57	0	6	123	440	885	1166	6684
Omaha	1355	1126	939	465	208	42	0	12	105	357	828	1175	6612
Scottsbluff	1231	1008	921	552	285	75	0	0	138	459	876	1128	6673
Valentine	1395	1176	1045	579	288	84	9	12	165	493	942	1237	7425
Nevada													
Elko	1314	1036	911	621	409	192	9	34	225	561	924	1197	7433
Ely	1308	1075	977	672	456	225	28	43	234	592	939	1184	7733
Las Vegas	688	487	335	111	6	0	0	0	0	78	387	617	2709
Reno	1073	823	729	510	357	189	43	87	204	490	801	1026	6332
Winnemucca	1172	916	837	573	363	153	0	34	210	536	876	1091	6761

New Hampshire													
Concord	1358	1184	1032	636	298	75	6	50	177	505	822	1240	7383
Mt. Wash. Obsy.	1820	1663	1652	1260	930	603	493	536	720	1057	1341	1742	13817
New Jersey													
Atlantic City	936	848	741	420	133	15	0	0	39	251	549	880	4812
Newark	983	876	729	381	118	0	0	0	30	248	573	921	4859
Trenton	989	885	753	399	121	12	0	0	57	264	576	924	4980
New Mexico													
Albuquerque	930	703	595	288	81	0	0	0	12	229	642	868	4348
Clayton	986	812	747	429	183	21	0	6	66	310	699	899	5158
Raton	1116	904	834	543	301	63	9	28	126	431	825	1048	6228
Roswell	840	641	481	201	31	0	0	0	18	202	573	806	3793
Silver City	791	605	518	261	87	0	0	0	6	183	525	729	3705
New York													
Albany	1311	1156	992	564	239	45	0	19	138	440	777	1194	6875
Binghamton (AP)	1277	1154	1045	645	313	99	22	65	201	471	810	1184	7286
Binghamton (PO)	1190	1081	949	543	229	45	0	28	141	406	732	1107	6451
Buffalo	1256	1145	1039	645	329	78	19	37	141	440	777	1156	7062
Central Park	986	885	760	408	118	9	0	0	30	233	540	902	4871
J. F. Kennedy Intl.	1029	935	815	480	167	12	0	0	36	248	564	933	5219
Laguardia	973	879	750	414	124	6	0	0	27	223	528	887	4811
Rochester	1234	1123	1014	597	279	48	9	31	126	415	747	1125	6748
Schenectady	1283	1131	970	543	211	30	0	22	123	422	756	1159	6650
Syracuse	1271	1140	1004	570	248	45	6	28	132	415	744	1153	6756
North Carolina													
Asheville	784	683	592	273	87	0	0	0	48	245	555	775	4042
Cape Hatteras	580	518	440	177	25	0	0	0	0	78	273	521	2612
Charlotte	691	582	481	156	22	0	0	0	6	124	438	691	3191
Greensboro	784	672	552	234	47	0	0	0	33	192	513	778	3805
Raleigh	725	616	487	180	34	0	0	0	21	164	450	716	3393
Wilmington	546	462	357	96	0	0	0	0	0	74	291	521	2347
Winston-Salem	753	652	524	207	37	0	0	0	21	171	483	747	3595

State/City	Jan	Feb	Mar	Apr	May	Jun	Jul	Aug	Sep	Oct	Nov	Dec	Year
North Dakota													
Bismarck	1708	1442	1203	645	329	117	34	28	222	577	1083	1463	8851
Devils Lake	1872	1579	1345	753	381	138	40	53	273	642	1191	1634	9901
Fargo	1789	1520	1262	690	332	99	28	37	219	574	1107	1569	9226
Williston	1758	1473	1262	681	357	141	31	43	261	601	1122	1513	9243
Ohio													
Akron	1138	1016	871	489	202	39	0	9	96	381	726	1070	6037
Cincinnati	970	837	701	336	118	9	0	0	54	248	612	921	4806
Cleveland	1159	1047	918	552	260	66	9	25	105	384	738	1088	6351
Columbus	1088	949	809	426	171	27	0	6	84	347	714	1039	5660
Dayton	1097	955	809	429	167	30	0	6	78	310	696	1045	5622
Mansfield	1169	1042	924	543	245	60	9	22	114	397	768	1110	6403
Sandusky	1107	991	868	495	198	36	0	6	66	313	684	1032	5796
Toledo	1200	1056	924	543	242	60	0	16	117	406	792	1138	6494
Youngstown	1169	1047	921	540	248	60	6	19	120	412	771	1104	6417
Oklahoma													
Oklahoma City	868	664	527	189	34	0	0	0	15	164	498	766	3725
Tulsa	893	683	539	213	47	0	0	0	18	158	522	787	3860
Oregon													
Astoria	753	622	636	480	363	231	146	130	210	375	561	679	5186
Burns	1246	988	856	570	366	177	12	37	210	515	867	1113	6957
Eugene	803	627	589	426	279	135	34	34	129	366	585	719	4726
Meacham	1209	1005	983	726	527	339	84	124	288	580	918	1091	7874
Medford	918	697	642	432	242	78	0	0	78	372	678	871	5008
Pendleton	1017	773	617	396	205	63	0	0	111	350	711	884	5127
Portland	825	644	586	396	245	105	25	28	114	335	597	735	4635
Roseburg	766	608	570	405	267	123	22	16	105	329	567	713	4491
Salem	822	647	611	417	273	144	37	31	111	338	594	729	4754
Sexton Summit	958	809	818	609	465	279	81	81	171	443	666	874	6254

Pennsylvania												
Allentown	1116	1002	849	471	167	24	0	0	90	353	693	1045	5810
Erie	1169	1081	973	585	288	60	0	25	102	391	714	1063	6451
Harrisburg	1045	907	766	396	124	12	0	0	63	298	648	992	5251
Philadelphia	1014	890	744	390	115	12	0	0	60	291	621	964	5101
Pittsburgh	1119	1002	874	480	195	39	0	9	105	375	726	1063	5987
Reading	1001	885	735	372	105	0	0	0	54	257	597	939	4945
Scranton	1156	1028	893	498	195	33	0	19	132	434	762	1104	6254
Williamsport	1122	1002	856	468	177	24	0	9	111	375	717	1073	5934
Rhode Island												
Block Island	1020	955	877	612	344	99	0	16	78	307	594	902	5804
Providence	1110	988	868	534	236	51	0	16	96	372	660	1023	5954
South Carolina												
Charleston	487	389	291	54	0	0	0	0	0	59	282	471	2033
Columbia	570	470	357	81	0	0	0	0	0	84	345	577	2484
Florence	552	459	347	84	0	0	0	0	0	78	315	552	2387
Greenville	648	535	434	120	12	0	0	0	0	112	387	636	2884
Spartanburg	663	560	453	144	25	0	0	0	15	130	417	667	3074
South Dakota												
Huron	1628	1355	1125	600	288	87	9	12	165	508	1014	1432	8223
Rapid City	1333	1145	1051	615	326	126	22	12	165	481	897	1172	7345
Sioux Falls	1544	1285	1082	573	270	78	19	25	168	462	972	1361	7839
Tennessee												
Bristol	828	700	598	261	68				51	236	573	828	4143
Chattanooga	722	577	453	150	25				18	143	468	698	3254
Knoxville	732	613	493	198	43				30	171	489	725	3494
Memphis	729	585	456	147	22				18	130	447	698	3232
Nashville	778	644	512	189	40				30	158	495	732	3578
Oak Ridge (CO)	778	669	552	228	56				39	192	531	772	3817

State/City	Jan	Feb	Mar	Apr	May	Jun	Jul	Aug	Sep	Oct	Nov	Dec	Year
Texas													
Abilene	642	470	347	114	0	0	0	0	0	99	366	586	2624
Amarillo	877	664	546	252	56	0	0	0	18	205	570	797	3985
Austin	468	325	223	51	0	0	0	0	0	31	225	388	1711
Brownsville	205	106	74	0	0	0	0	0	0	0	66	149	600
Corpus Christi	291	174	109	0	0	0	0	0	0	0	120	220	914
Dallas	601	440	319	90	6	0	0	0	0	62	321	524	2363
El Paso	685	445	319	105	0	0	0	0	0	84	414	648	2700
Fort Worth	614	448	319	99	0	0	0	0	0	65	324	536	2405
Galveston	350	258	189	30	0	0	0	0	0	0	138	270	1235
Houston	384	288	192	36	0	0	0	0	0	6	183	307	1396
Laredo	267	134	74	0	0	0	0	0	0	0	105	217	797
Lubbock	800	613	484	201	31	0	0	0	18	174	513	744	3578
Midland	651	468	322	90	0	0	0	0	0	87	381	592	2591
Port Arthur	384	274	192	39	0	0	0	0	0	22	207	329	1447
San Angelo	567	412	288	66	0	0	0	0	0	68	318	536	2255
San Antonio	428	286	195	39	0	0	0	0	0	31	207	363	1549
Victoria	344	230	152	21	0	0	0	0	0	6	150	270	1173
Waco	536	389	270	66	0	0	0	0	0	43	270	456	2030
Wichita Falls	698	518	378	120	6	0	0	0	0	99	381	632	2832
Utah													
Milford	1252	988	822	519		87	0	0	99	443	867	1141	6497
Salt Lake City	1172	910	763	459		84	0	0	81	419	849	1082	6052
Wendover	1178	902	729	408	177	51	0	0	48	372	822	1091	5778
Vermont													
Burlington	1513	1333	1187	714	353	90	28	65	207	539	891	1349	8269

Virginia													
Cape Henry	694	633	536	246	53	0	0	0	0	112	360	645	3279
Lynchburg	849	731	605	267	78	0	0	51	51	223	540	822	4166
Norfolk	738	655	533	216	37	0	0	0	0	136	408	698	3421
Richmond	815	703	546	219	53	0	0	0	36	214	495	784	3865
Roanoke	834	722	614	261	65	0	0	0	51	229	549	825	4150
Wash. Nat'l A. P.	871	762	626	288	74	0	0	0	33	217	519	834	4224
Washington													
Olympia	834	675	645	450	307	177	68	71	198	422	636	753	5236
Seattle	738	599	577	396	242	117	50	47	129	329	543	657	4424
Seattle Boeing	831	655	608	411	242	99	34	40	147	384	624	763	4838
Seattle Tacoma	822	678	657	474	295	159	56	62	162	391	633	750	5145
Spokane	1231	980	834	531	288	135	9	25	168	493	879	1082	6655
Stampede Pass	1287	1075	1085	855	654	483	273	291	393	701	1008	1178	9283
Tatoosh Island	713	613	645	525	431	333	295	279	306	406	534	639	5719
Walla Walla	986	745	589	342	177	45	0	0	87	310	681	843	4805
Yakima	1163	868	713	435	220	69	0	12	144	450	828	1039	5941
West Virginia													
Charleston	880	770	648	300	96	9	0	0	63	254	591	865	4476
Elkins	1008	896	791	444	198	48	9	25	135	400	729	992	5675
Huntington	880	764	636	294	99	12	0	0	63	257	585	856	4446
Parkersburg	942	826	691	339	115	6	0	0	60	264	606	905	4754
Wisconsin													
Green Bay	1494	1313	1141	654	335	99	28	50	174	484	924	1333	8029
La Crosse	1504	1277	1070	540	245	69	12	19	153	437	924	1339	7589
Madison	1473	1274	1113	618	310	102	25	40	174	474	930	1330	7863
Milwaukee	1376	1193	1054	642	372	135	43	47	174	471	876	1252	7635
Wyoming													
Casper	1290	1084	1020	657	381	129	6	16	192	524	942	1169	7410
Cheyenne	1228	1056	1011	672	381	102	19	31	210	543	924	1101	7278
Lander	1417	1145	1017	654	381	153	6	19	204	555	1020	1299	7870
Sherian	1355	1154	1054	642	366	150	25	31	219	539	948	1200	7683

found by multiplying the overall heat loss coefficient U (in BTU/hr°F) times 24 hours/day times the number of degree-days in the month.

The conventional approach to calculating the maximum heating demand for a structure is to take an outside design temperature, such as 10°F, and calculate the maximum heat input (in BTU/hr) required to maintain the interior to 70°F, assuming no other internal heat generation. The overall heat loss coefficient U is this maximum heat input divided by the difference between 70° and the outside design temperature. For example, if an HVAC contractor has sized a house for a 75,000 BTU/hr furnace based on an outside design temperature of $-5°F$, then the overall U factor for the house is found by dividing 75,000 by $70 - (-5)$; therefore U is 1000 BTU/hr°F. Thus, from Table 7, the total heating required for this house in Mobile, Alabama in February is $1000 \times 24 \times 300$, or 7,200,000 BTU.

One refinement to this formula has been suggested by the electrical industry. Since electrically (and solar) heated structures have no fuel-air draft requirements and are usually constructed with vapor barriers, the heating requirements are not as great as with gas- and oil-heated facilities, for which the degree-day formula was originally developed. Data taken with electrical houses indicate that the heating requirement is reduced to 77% of that predicted by the degree-day formula, so the monthly heating requirement for this type of structure as calculated above should be further multiplied by 0.77.

The domestic hot water requirement per person is generally considered to be 20 gal/day of 140°F hot water. Thus, the BTU's used per day is 20 gal/day times the number of family members times 8.345 lb/gal times the difference between 140° and the inlet temperature. The inlet temperature depends on location, but is usually in the neighborhood of 40°F. For example, a family of five living in an area with 40°F inlet temperature would require $20 \times 5 \times 8.345 \times (140 - 40)$, or 83,450 BTU/day for heating domestic hot water. The month-by-month inlet water temperatures are usually available from the local water works.

Flat Plate Collector Systems

A typical solar heating system employing a flat plate collector is illustrated by Figure 12. A flat plate collector located on a southward sloping roof heats water which circulates through a coil in the hot water tank, then through a coil in a large warm water tank before being returned to the collector. In most areas of the country the heat transfer fluid flowing through the collector should be an anti-freeze solution to prevent freezing of the fluid in the collector tubes in the winter. The system shown in Figure 12 provides for two levels of heat storage; the hottest water which is stored in the hot water tank is used for building services, and the warm water in the large tank heats water circulating through pipes in the house. The heat reservoir for a single dwelling could be a tank 6 feet in diameter, 5 feet deep, and insulated on all sides. An auxiliary heating system is necessary to provide heat during extended cold cloudy periods when the supply of solar heat is not adequate.

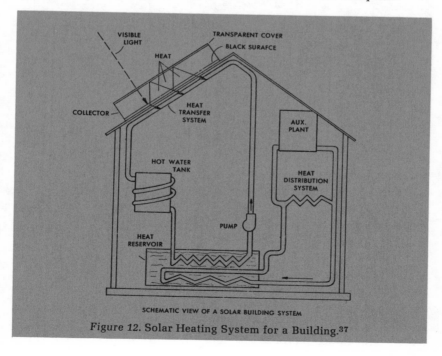

SCHEMATIC VIEW OF A SOLAR BUILDING SYSTEM

Figure 12. Solar Heating System for a Building.[37]

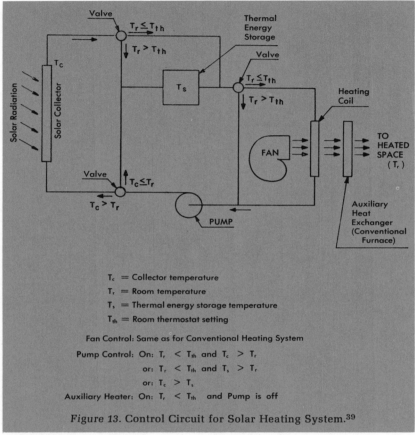

Figure 13. Control Circuit for Solar Heating System.[39]

Several studies have been conducted to determine optimal control systems for solar home heating systems, such as the one illustrated by Figure 13. The main object of the control system is to extract heat from the solar collector when it is available, but to shut off the flow through the collector whenever the collector temperature drops below the storage temperature. In this system a separate auxiliary heater is provided. The pump circulates water through the collector whenever the collector outlet temperature exceeds the storage temperature. If the room temperature is lower than both the collector temperature and the thermostat setting, water from the collector is circulated directly through the heating pipes in the house. If the room temperature is lower than both the thermostat setting and the storage temperature, but higher than the collector temperature (such as at night), hot water from the storage tank is circulated through the room. Thus, the solar heat is transferred directly to the room if it is too cool, and transferred to the storage tank for later use if the room is already warm enough. This is a fairly standard type of solar thermal control system using a single water pump and three valves. The hot water heat exchanger for heating air (similar to an automobile radiator) can be installed in a conventional forced air furnace.

Flat plate collectors are also used for directly heating air for house heating. Water is usually used because of the simple storage system, which is just an insulated tank. The simplest air heater is a flat black plate covered by a transparent sheet, with air flowing in the gap between. However, higher temperatures are achieved if the air flows through or beneath the black absorbing surface, and the air gap beneath the transparent cover and plate is stagnant. A good collecting surface is a V-corrugated absorber plate with a spectrally selective coating (absorptivity 0.80 in the visible, 0.05 in the infrared). Absorbers of this type heated air to 170°F with 40% collection efficiency for an insolation of 160 BTU/ft²/hr and an ambient dry bulb temperature of 74.6°F. For an insolation of 300 BTU/ft²/hr a temperature of 210°F was reached with 40% collection efficiency.[40] The maximum temperature of the air can be increased an additional 10–15°F with no loss in collector efficiency by allowing the air to flow over the absorbing surface and then back under the absorber (2 passes) instead of the standard single-pass configuration.[41]

Solar Heating System Design and Installation

A solar heating and domestic hot water system can be installed by any individual sufficiently skilled in carpentry, plumbing and electrical work to add a room, finish a basement, or perform other similar construction and repairs. The required electrical work involves some low-voltage wiring; the pumps and control units can be plugged into existing electrical outlets if desired. Plumbing techniques are no more sophisticated than those required to install a sink. The solar collectors can be built into the roof or mounted on

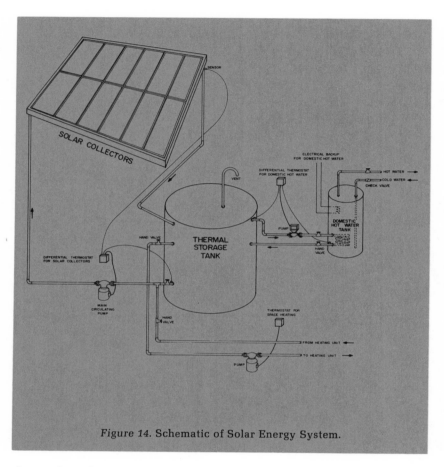

Figure 14. Schematic of Solar Energy System.

the roof, or located on the ground, over a patio or any other convenient location that is unshaded most of the day.

The system described here uses water to transfer and store heat. The water is heated in the solar collectors and the heated water is pumped to the thermal storage tank for later use, or to the home for heating. A copper coil in a hot water storage tank heats domestic hot water, as needed, 24 hours a day. This particular system is designed so that the collectors will drain completely of water whenever the main circulating pump is off; this prevents freezing in the collectors and improves the performance of the system by reducing the thermal inertia of the collectors, allowing the collectors to reach operating temperature earlier in the day. Since ordinary tap water is used as the circulating fluid, and the collectors are drained daily, all piping in the collectors and throughout this system must be of materials not subject to corrosion, such as copper or stainless steel.

A first step in designing a solar system is to contact the major solar collector manufacturers to obtain their latest specifications and prices. In addition to the list of manufacturers of liquid heating collectors presented in the Appendix, you should also refer to your local yellow pages under "Solar Energy." A word of caution: Before buying, examine the collector specifications, installed collectors if any are nearby, the reputation of the manufacturer, and the warranty. Also try to get independent verification of performance claims.

Once you have examined the various solar collector options and have selected a collector that is attractive, durable, and offers the most heat at the least cost, you can then proceed to size the array. This requires estimating the collector area required to provide the percentage of the load to be supplied with solar energy. In order to provide around 70% of the heating requirements of a well-insulated home or building with 50%-efficient collectors, the collector area required is usually in the range of one square foot of collector for every three or four square feet of heated floor space; to deliver about 50% of the heat, one square foot of collector is needed for every five to six square feet of floor space. If the overall average efficiency (not the instantaneous efficiency) of the collectors is different from 50%, the collector area required is correspondingly smaller or larger.

The collector efficiency value you needed to calculate the heat production by the array is the long-term average value determined by averaging months of experimental data or by a proper numerical analysis utilizing instantaneous values and meteorological parameters. The long-term average efficiency value should be available from the collector manufacturer. The manufacturer should tell you how the value was obtained and who performed the measurements —themselves, an independent laboratory, or a government agency.

In order to estimate the heat output by month, use Table 1 to find the average daily solar radiation on a horizontal surface for the city closest to your home (values given in BTU per day per square foot of horizontal surface). Choose the collector slope based on architectural considerations and the fact that, for maximum heat delivery, the best slope is to the south and about 10 degrees greater than the latitude; that is, at 45 degrees north latitude, the best slope for heating is about 55 degrees (from the horizontal), facing southward. Refer then to Table 2 which lists the daily solar radiation incident on a southward-facing surface as a function of month, latitude and slope. Some interpolation may be required to obtain the values for latitude and slope. If the slope is 10 degrees greater than the latitude, use the L+10 column. Then, for each month, divide the value for this slope by the horizontal value from Table 2 and multiply the result by the values for your location from Table 1. You now have estimated the daily solar radiation, in BTU per square foot per day, falling on the solar collectors. You can carry out this exercise for different collector slopes to determine the effects of other slope angles.

Now that you have estimated the average amount of solar radiation falling on your collectors, you can calculate the heat production by multiplying by the long-term average efficiency value and the total collector area. It is important to note whether the efficiency value is based on gross collector area (based on outside dimensions) or aperture area (based on area of absorbing surface exposed to sunlight); multiply by the area appropriate to the efficiency value. Since the heat production is proportional to the collector area, you can now determine what area you need to supply the amount of heat you need. Usually the most cost-effective system is one sized to supply between 60% and 70% of the yearly heat demand.

Now that you have selected your collector type, slope and array size, you can proceed to design your system. Your thermal storage tank should be steel, coated with coal tar epoxy properly applied, or another material and coating that will last 30 years with wet temperatures sometimes approaching boiling. In sunny climates with little cloud cover, the best thermal storage capacity is usually around 1.5 gallons per square foot of collector; in more typical climates with frequent cloud cover, about 3 gallons per square foot of collector is recommended. If the collector area is greater than one-third the heated floor space, you could add another couple of gallons per square foot of collector area. Call your local steel fabricator or tank manufacturer to get the lowest cost tank in the size range you need. A final word of caution: If you use any tank material other than steel, be sure that it will remain watertight and withstand repeated thermal cycling over the lifetime of the solar system.

This particular heating system is designed so that the collectors drain whenever the main circulating pump is off; therefore the collectors must be higher than the tank, and the water line from the collector outlet manifold must always slope downward. The part of the water line leading to the collectors which is exposed to freezing temperatures must also slope downward and be higher than the surface of the water in the thermal storage tank. The thermal storage tank is never completely filled, but has an air gap at the top of sufficient volume to accommodate thermal expansion of the water (about 4% of the total water volume) plus enough volume to accommodate the water draining from the collectors. The return line from the collectors enters the top of the tank (above the water level) and the line to the collectors exits near the bottom of the tank. The main circulating pump must be located lower than the level of the water surface in the thermal storage tank, so that it will remain primed.

If you want the pipes to enter the sides of the tank, the manufacturer should install pipe fittings. There are six required in all: one at the top for the water pipe returning from the collectors; one near the bottom for the water pipe to the collectors (water returning from the home heating unit enters this pipe through a T between the tank and pump); one near the top but below the tank water level for the pipe going to the home heating unit; two near the top but below the water surface for domestic water heating; and one at the top for a vent line. Except for these penetrations, the tank should be airtight. A level gauge or float valve can be added to ensure that the tank is properly filled. The vent line to the outside is necessary to maintain the tank pressure at atmospheric.

A small pump circulates water through the copper coil in the bottom of the potable water tank, and back to the thermal storage tank. The thermal storage tank must be well insulated—the more insulation the better. The same is true for the piping. If the tank is above-grade and dry, use about 12 inches of fiberglass (or equivalent) surrounding the tank. If the tank is underground, the tank can be surrounded first by dry gravel, then slabs of outdoor styrofoam, then more gravel. Dry gravel is a fairly good insulator and provides some thermal storage. Be sure the bottom of the tank is above the water table, and the top insulation is covered to keep water out of the gravel. If the tank is below-grade, provision must be made to keep the main circulating pump below the level of the water in the tank.

The collector manifolds should have a slope of about 1/8 inch to the foot or more to be sure that they drain when the pump is off. Likewise, connect the collectors so that they all drain completely.

Now that you have located the collector, storage tank and other system components, choose a reasonable diameter for the piping. The potable water lines, for example, could be 3/4-in. and the main lines to/from the collectors 1 in. Considering a flow rate of 1/40 gal/minute for each square foot of collector, divide the collector array area by 40 to get the flow rate in gpm. Then using Tables 8 and 9, calculate the pressure drop in feet of head (divide by 0.4335 if you prefer working in psi). For example, using Table 8, at a flow rate of 8 gpm through 3/4-in. Type L copper tubing, the pressure drop is 17 feet of head for each 100 feet of tubing. Add the pressure drops due to elbows, bends, etc. by adding the "equivalent feet of pipe" from Table 9 to the length of run before calculating the pressure drop from Table 8.

Having calculated the pressure drop throughout a complete flow circuit, select a pump that will provide the required pressure head at the desired flow rate. The main collector circulating pump must also provide the necessary head at a low flow rate to raise the water to the top of the collectors when the pump is first turned on.

For controls, two differential thermostats are needed—one for the potable water pump and one for the collector pump. These can be purchased for about $50 each and should be installed in accordance with the instructions provided with the units. One differential thermostat has sensors attached to the copper pipes at the collector

Table 8. Pressure Drop in Copper Pipe (ft of head)

Flow (gpm)	3/8	1/2	5/8	3/4	1	1 1/4	1 1/2	2	2 1/2	3	3 1/2	4	5	6
					Head Loss/100 ft – Diameter of Type L Copper Tubing (in.)									
1	8	3	1											
2	28	9	4	2										
3	55	18	7	3										
4	92	30	11	5	1									
5	140	44	16	8	1									
6	190	60	23	11	3	1								
7	240	81	28	14	4	1								
8	300	97	37	17	5	2								
9	390	120	46	21	6	2								
10	460	150	58	25	7	3	1							
15		300	115	50	15	5	2							
20			200	90	23	9	4	1						
25			280	130	37	13	6	2						
30				180	51	17	8	2						
35				230	65	23	11	3	1					
40					81	30	13	4	1					
45					105	35	16	4	2					
50					127	44	20	5	2					
60						60	25	7	3	1				
70						76	35	9	4	1				
80						92	44	12	4	2				
90							55	14	5	2	1			
100							67	17	7	3	1			
150								35	13	6	3	1		
200								60	21	9	5	3		
250									32	14	7	4	1	
300									44	20	9	5	2	
350										25	12	7	2	
400										32	16	8	2	
450										37	18	10	4	2
500											23	12	4	2
600											32	16	6	3
700												21	8	3
800												23	9	4
900													12	5
1000													14	6

Table 9. Pressure Drop across Copper Fittings

Nominal Size of Tube (in.)	90° Elbow	45° Elbow	Tee, Run	Tee, Side Outlet	90° Bend	180° Bend
			Wrought-Copper Fittings			
3/8	1/2	1/2	1/2	1	1/2	1/2
1/2	1/2	1/2	1/2	1	1/2	1
5/8	1/2	1/2	1/2	2	1	1
3/4	1	1/2	1/2	2	1	2
1	1	1	1/2	3	2	2
1 1/4	2	1	1/2	4	2	3
1 1/2	2	2	1	5	2	4
2	2	2	1	7	3	8
2 1/2	2	3	2	9	4	16
3	3	4			5	20
3 1/2	4				7	24
4					8	28
5					10	37
6					13	47

exit and at the exit of the thermal storage tank, so that the main circulating pump is turned on whenever the collector temperature exceeds the temperature of the water near the bottom of the thermal storage tank. Likewise, the other differential thermostat has sensors on the copper pipes where the potable water line exits the thermal storage tank and at the exit of the potable water tank, so that the potable water circulating pump is turned on whenever the collector temperature exceeds the temperature of the water near the bottom of the potable water tank. The pump circulating hot water from the tank to the home-heating systems is activated by the standard home thermostat.

This basic solar heating and hot water system (Figure 14) can be expanded for operating a solar-assisted heat pump for auxiliary heating and summer air conditioning (Figure 15), and for solar air conditioning (Figure 16). As shown in Figure 15, a coil of copper pipe in the thermal storage tank permits a glycol solution to be circulated through this coil to the heat pump. Glycol solution must be used in this loop to prevent freezing in the outside fan-coil unit in the winter. Normally, the home is heated by circulating solar-heated water from the thermal storage tank directly through the coil in the return duct. Whenever the available solar energy is insufficient for direct heating, the heat pump operates using solar-heated glycol solution as the source of heat. The heat pump operates in reverse to cool in the summer. Figure 17 shows a solar home in Georgia that uses this system.

At least one home-sized water-fired chiller for solar systems is now on the market (early 1977)—the Arkla-Servel SOLAIRE-36, which is rated at three tons with 195°F firing water. For cooling (Figure 16), solar-heated water is circulated through the generator section of the chiller, and 40°F chilled water is produced in the evaporator section and circulated to a chilled water storage tank. Condensing water, for waste heat rejection, can come from a cooling tower, an exterior fan-coil unit, a swimming pool, or any other convenient source. The cooler the condensing water, the better the operation of the chiller. Auxiliary energy for heating and cooling is provided by a boiler, which can use gas, oil, coal, or even wood, depending on the design of the boiler. Either hot or cold water is

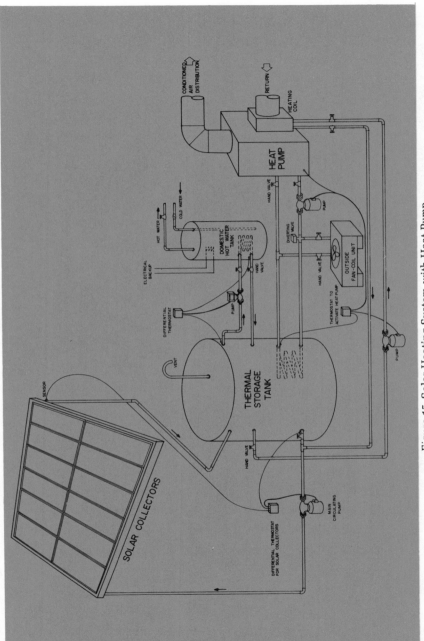

Figure 15. Solar Heating System with Heat Pump.

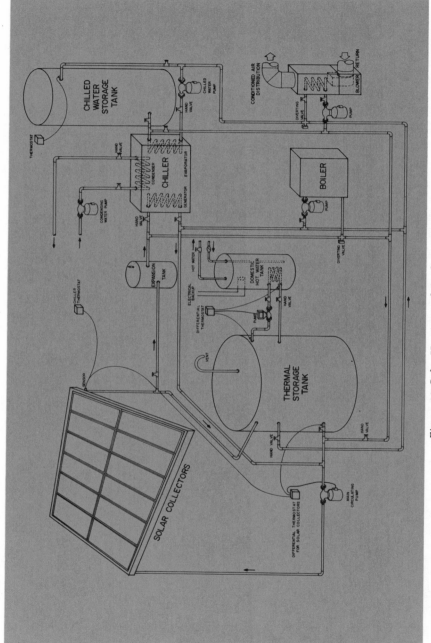

Figure 16. Solar Heating and Cooling System.

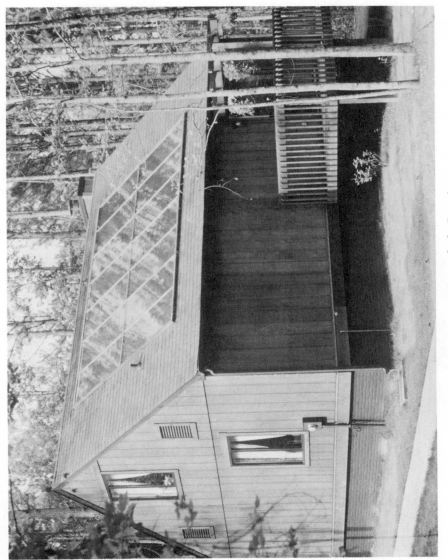

Figure 17. Solar Home in Shenandoah, Georgia.

Figure 18. Solar Domestic Hot Water System with Hexagonal Skylight Collectors, in Atlanta.

circulated through the heat exchanger coil in the distribution system to provide either heating or cooling. Whenever a glycol solution is used in a flow loop, the flow rate and pump head must be increased, as illustrated by Figure 19 for a 50% glycol solution.

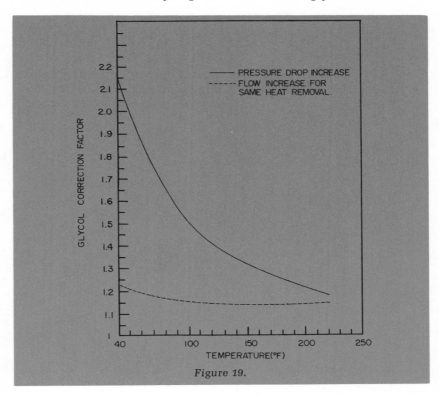

Figure 19.

Thermal Storage

The most common heat transfer fluid for a solar system is water, and the easiest way to store thermal energy is by storing the water directly in a well-insulated tank. The optimum tank size for flat plate collector system is usually about 2 gal/ft². If air is used as the heat transfer medium, it can be circulated directly through a gravel tank for heat storage, then brought from the gravel tank to the house, as required, for heating and other applications. More compact heat storage is possible with phase-change materials such as Glaubers' salt (sodium sulfate decahydrate, $Na_2SO_2\text{-}10H_2O$).

Table 10. Thermal Storage of One Million BTU with 20°F Temperature Change

	Water	Rocks	Phase Change Material
Specific heat (BTU/lb°F)	1.0	0.2	0.5
Heat of fusion (BTU/lb)	—	—	100
Density (lb/ft³)	62	140	100
Weight (lb)	50,000	250,000	10,000
Volume (ft³) with 25% passage	1,000	2,150	125

Water, rocks and a typical phase-change material may be compared as shown in Table 10.[42] Water can store heat over a range of temperatures approaching 200°F, and rocks can store heat (or coolness) at any practical temperature, but phase-change materials melt and solidify at one temperature. Thus rocks and water can store heat in the winter and "coolness" in the summer, whereas two separate salt systems would be required to accomplish this. Phase-change materials also cost more per BTU of heat storage than water or rocks; gravel costs are about $5.00/ton. The advantages of the phase-change material are, of course, considerably reduced weight and volume.

Roof Ponds

A simple technique for heating and cooling a house is to locate a pond of water 6–10 inches deep on the roof. The pond is covered by thermal insulating panels which can be opened or closed. In the winter all the water is enclosed in polyethylene bags atop a black plastic liner. Sunlight heats the water to about 85°F during the day. At night, the insulating panels cover the water to prevent loss of the heat to space. During the summer, the insulating panels are open at night and closed during the day, so the water is cooled by radiation to space at night. Tests with a small 10 ft × 12 ft structure in Phoenix, Arizona, showed that temperatures were maintained close to 70°F year round by pulling a rope twice a day, even though ambient temperatures ranged from subfreezing to 115°F.[43] A new two-bedroom house with a 10-inch-deep roof pond is now being tested (Figure 20) in California at Atascadero near Paso Robles which has recorded temperature extremes of 10°F and 117°F.[44] The horizontal roof collector is not expected to meet the full heat demand because ambient air temperatures are lower, cloud

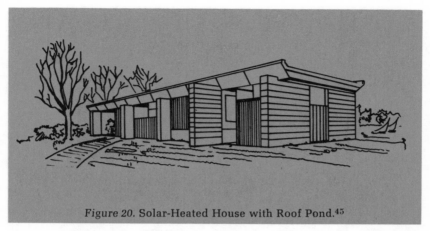

Figure 20. Solar-Heated House with Roof Pond.[45]

cover is greater, and the location is two degrees more northerly than the Phoenix location of the test room. Summer cooling, however, should be better than at the Phoenix location. The roof pond is not visible at ground level.

The French System

An economically attractive natural circulation solar heating system has been developed in France and tested with several full-sized houses. A south-facing concrete wall is painted black and covered with glass, with an air gap between the wall and the glass. Sunlight heats the concrete wall and air circulates through vents from the room, rises as it is heated in the air gap, and flows out into the room through vents near the ceiling. The concrete wall acts both as heat collector and heat storage medium, since the warmed wall continues to provide heat in the evening. During the summer the room vent near the ceiling is closed and an exterior vent opened, so the system operates as before but the warmed air flows outside, drawing cooler air into the room from the north side of the house. This type of solar heating system can supply half or more of the year-round heating requirements of these homes in France.

Costs

The present range of system costs for solar heat is from less than $2.00 per million BTU (MBTU) to about $5.00 per MBTU, depending on

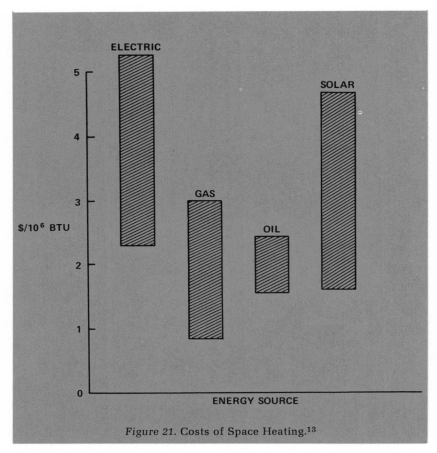

Figure 21. Costs of Space Heating.[13]

climate, the type of system used, and prevailing materials and labor costs. As seen from Figure 21, solar heating costs tend to compare favorably with electric heat, but unfavorably with natural gas. As prices of oil and gas rise and the technology for solar heating advances, solar heating systems will become increasingly competitive.

Existing Solar Homes and Buildings

Tables 11–13 summarize data for some solar-heated units. The solar-heated units are presented in a past/present format and a future format. Only those buildings that have had some perform-

Table 11. Solar Houses and Buildings with Water Storage

Solar House	Year	Floor Area	Collector	Store	Cooling	Solar Heat Supplied
Skytherm, Phoenix	1967	120 ft^2	170 ft^2/water bags	same	Rooftop, night	All
Skytherm*, Cal.	1973	1140 ft^2	304 ft^2/water bags	same	Rooftop, night	All/carry-over, 6 days
MIT III, Mass.	1949	600 ft^2	400 ft^2/water type	1500 gal	—	75-85%/carry-over 2 days
MIT IV, Mass.	1959	1450 ft^2	640 ft^2/water type 60° tilt	1500 gal	—	57%
Thomason* No. 1	1959	1036 ft^2	836 ft^2/water type	1600 gal - 50 tons stones	Rooftop, night	95% first year
Washington D.C. No. 2	1960	1008 ft^2	560 ft^2/water type	1600 gal - 60 tons stones	3/4 hp, refrig.	70-85% with reflector
Washington D.C. No. 3	1965	3256 ft^2	960 ft^2/water type	3000 gal - 25 tons stones	3/4 hp, refrig.	90% with reflector
Ouroboros, Minn.	1974	2000 ft^2	600 ft^2/double-glazed 45° tilt/500 ft^2 –wall	2000 gal	—	80%
Mathew, Oregon	1967	1650 ft^2	725 ft^2/82° tilt/ water type	8000 gal	—	80-100% with reflector
Brisbane*, Australia	1966	1318 ft^2	720 ft^2/water type	water–rockpile	LiBr absorption	All
Swedish, Italy	1960	1940 ft^2	320 ft^2/double-glazed on wall	800 gal	reverse cycle	50-70%
French*	1967	300 m^2	48 m^2/wall, air type	wall	reverse cycle	66%
Saunders, Mass.	1960	2600 ft^2	400 ft^2/wall double-glazed 1200 ft^2/air type	wall–thick floor	—	60%–both 40%–wall only

Table 11. Continued

Solar House	Year	Floor Area	Collector	Store	Cooling	Solar Heat Supplied
Baer*, N.M.	1971	2000 ft^2	180 ft^2/wall	90-55 gal barrels	–	90%
Meinel, Arizona	1973	2000 ft^2	300 ft^2/double-glazed wall	1000-1 gal polyethylene bottles	reverse cycle	30%
N. View H. S. Minn.	1974	16600 ft^2	5000 ft^2/double-glazed 55° tilt/water type	3000 gal	–	60%
B.P.H.* Office Bldg N. M.	1956	4300 ft^2	750 ft^2/single-glazed 60° tilt/water type	6000 gal	Heat pump	75%
Mass.* Audubon Society Bldg.	1973(?)	8000 ft^2	3500 ft^2/double-glazed air type		LiBr–15 ton	65-85%

Table 12. Solar Houses with PCM Storage

Solar House	Year	Floor Area	Collector	Store	Cooling	Solar Heat Supplied
State College,* N.M.	1953	1100 ft^2	460 ft^2/air type 45° tilt	Glauber salt 2 tons	–	50%
Peabody,* Mass.	1949	1250 ft^2	720 ft^2/air type double-glazed	Glauber salt 21 tons	–	80%, 6-day carry-over
Solar One*	1973	1300 ft^2	720 ft^2/single-glazed photovoltaic—45° tilt/ 144 ft^2/double-glazed wall/air type	8000 lb/$Na_2S_2O_3$ 5 H_2O, 1400 lb/ Na_2SO_4 10 H_2O 2600 lb/"coolness store" Eutectic salt formation	Heat Pump	90%, 2-day carry-over

Table 13. Recent Solar Houses and Buildings

Solar House	Year	Floor Area	Collector	Store	Cooling	Solar Heat Supplied
Colorado State College*	1974	3000 ft^2	760 ft^2/double-glazed Roll-bond/45° tilt	1100 gal	LiBr	65%
Pinchot, Conn.	1974	750 ft^2	450 ft^2/double-glazed selective coating/ 45° tilt	1500 gal		50-60%, 2-day carry-over
Barber, Conn.	1974	1900 ft^2	450 ft^2/double-glazed selective coating/ 45° tilt	1500 gal	refrig, cools tank	55%, 2-day carry-over
Ohio State College	1974	2200 ft^2	800 ft^2/double-glazed 45° tilt	4000 gal	LiBr	50-70%
Cary Bldg., N.Y.	1975	35,000 ft^2	5000 ft^2/double-glazed			50-70%
Fed. Office Bldg., N.H.	1976	131,000 ft^2	6000 ft^2/variable tilt 0°-70°	30,000 gal		
Wilson, W. Va.	1975	1400 ft^2	600 ft^2/water type	2800 gal	Heat pump	2-day carry-over
Grover Cleveland School, Boston	1974		4500 ft^2/Roll-Bond/ double-glazed 45° tilt	2000 gal	—	20%

ance data are shown. The past/present buildings were subdivided into categories based on their particular storage systems, water, rock, or PCM (Phase Change Material).

Analysis of Solar Heating and Cooling

Tybout and Lof[46] described six design parameters in their study of optimization (cost minimization) of a solar home. They varied each parameter in a computer analysis while keeping the other parameters fixed at three or four different values. One conclusion was that two parameters, the heat transfer coefficient of the storage tank insulation and the heat capacity of the collector, have practically no effect on system performance. They concluded that within the common practical ranges these two parameters can be neglected in cost minimization.

The findings on the other four parameters were as one would suspect. The collector tilt angle affects the total solar incidence on the collector; therefore, an optimum tilt angle can be found for a given location. Relatively high angles, 40°–70°, for a 40° north latitude location (the mean latitude for the United States population) were optimum.

The number of glass plates on the collector affect the optical and thermal losses. There is a trade-off between the two. It was shown that one or two plates of glass is optimal, one plate where thermal loss is low (in the South or Southwest) and two plates where thermal loss is high (in the North). Two plates should be used in the middle latitude section of the country.

Collector size for minimum solar heat cost for a 25,000 BTU/degree-day house in six locations was found to range from 208 ft^2 (Charleston, S.C.) to 521 ft^2 (Omaha, Nebraska), corresponding to 55% of the respective annual heating loads. In Santa Maria, California a 261 ft^2 collector can supply 75% of the annual heat requirement. In most situations the cost of solar heat near optimum levels is rather insensitive to collector size and the corresponding fraction of load carried. Costs rise sharply, however, if designs are based on carrying large fractions (over 90%) of the load. In structures with small or larger heat demands than 25,000 BTU/DD, optimum collector size is approximately proportional to the demand parameter.

Heat storage capacity for minimum solar heating cost in nearly all practical situations is 4.5 kg to 6.8 kg (10 to 15 lb) of water (or its thermal equivalent) per square foot of collector. This is the equivalent to one to two days' average winter heating requirement. Solar heating cost is not very sensitive to storage capacity in this general range.

Several computer programs are now in use to evaluate the performance of heating and cooling systems for buildings. The detailed engineering design information required to make accurate predictions for using solar energy include the effects of variations in flat plate collector designs (angle of tilt, number and type of cover plates, efficiencies, flow rates), storage capacity and form of storage (water, salts, solids, paraffins), load patterns (meteorological effects, use patterns, internal heat sources, temperature zones), building variables (orientation, window area, construction details, architectural features, shading, infiltration, heating and cooling distribution system options, etc.), and an economic analysis (sensitivity to first cost, comparisons of life cycle costs as functions of interest rates and fuel prices, variations between types of fuel, distributions with respect to location, etc.). In order to study the effects of each variable and select a successful path through the labyrinth of options, one must have an extremely versatile and relatively sophisticated computational tool.

The Post Office Department funded a detailed computer study of the energy utilization in postal facilities. The initial work is described in four volumes. Their investment spawned a selection of variations and improvements on the original program.

ASHRAE certified that the calculations provide reasonable estimates of the heating and cooling loads as well as costs of operating the building. The program calculates the heating and cooling requirements as functions of building construction details and can study variations in duct design and comfort conditioning control. Modified and expanded versions of this program are available at GATX, Niles, Illinois; Hittman Associates, Columbia, Md.; and at Consultants Computation Bureau, San Francisco. The Department of Housing and Urban Development is funding the use of this program for some of its studies.

The program does not treat the design of solar collectors and heat storage apparatus. It is primarily a building load study program. However, the effects on the heating and cooling load of adding collectors on the roof or walls can be calculated. Also, the use of recessed windows and other architectural features can be studied in detail. Infiltration heating and cooling loads are treated in the main text of the program.

The University of Wisconsin received a National Science Foundation (NSF) grant titled "Modeling of Solar Heating and Air Conditioning" beginning in the summer of 1972. One result of that contract is a computer program called TRNSYS. Models for a collector, pump, and control strategy can be studied as a combination.

Building loads can be derived from a National Research Council of Canada program, an ASHRAE program, and routines written at CSU. U.S. Weather Bureau tapes are used for meteorological input data. Output is available on X-Y plotters, CRT's, printers, or microfiche. The program is presently being used to study the dynamics of a proposed 417,000 ft^2 office building.

The Jet Propulsion Laboratory of the California Institute of Technology has a computer program for simulating hot water heating using solar energy. The Southern California Gas Company and NSF/RANN funded a small effort to explore possible effects of gas demand if solar water heaters were to be used extensively in the Los Angeles basin. The calculations performed apply to a 10-unit apartment building.

Los Alamos Scientific Laboratory (LASL) and the Sandia Corporation in New Mexico developed computer models to study solar communities. Subdivisions or groups of housing having utility services derived from solar energy are examined. Studies of energy collection at either individual houses or a central park are being performed. Programs for modeling collectors, heat storage apparatus, and fluid transmission lines have been developed. Cost sensitivity analyses and comparisons with the cost of gas, coal and electrical energy have been performed.

Chapter 5

SOLAR WATER HEATERS

Solar water heaters are currently in widespread use throughout many sunny areas of the world. A common arrangement is to have a flat plate solar collector on the roof that provides hot water by natural circulation to a tank located higher on the roof. The roof tank can be designed to look like a chimney or located in the attic. In Japan there are about 2½ million solar water heaters of several different types in use.[47] The Japanese units employ a storage tank and collector as an integral unit, whereas in other countries the storage tank is usually separated. The simplest and oldest type is a flat open tank on the roof (about $10 with a black bottom) which supplies water at 130°F in the summer and as high as 80°F in the winter. Since the water is sometimes contaminated by dust, a polyethylene film covering the tank can be added for a few extra dollars. The transparent cover lasts about three years, and increases the water temperature as well as preventing contamination. The standard heater size is about 3 ft wide, 6 ft long and 5 in. deep. These flat tank water heaters are cheap, but suffer a major disadvantage because they must be mounted horizontally, so they are not very effective in the winter when the sun is low. Closed pipe collectors can be mounted at a more optimum angle to the sun and thus provide hotter water during the winter months. The pipes are made of glass, plastic or stainless steel painted black and mounted in a frame covered with glass or transparent polyethylene plastic.

In the United States about 60–70 ft² of collector can supply 75% of the water heating needs of apartments. One study called Project SAGE (Solar Assisted Gas Energy)[38] in southern California, a joint project of the Jet Propulsion Laboratory and the California Gas Company, studied the technical and economic aspects of a solar-assisted gas and electric water heating system for a typical southern California apartment building. Figure 22 illustrates a solar-electric hot water system for an apartment complex, with a single collector and storage tank, which reduces the cost of collecting the solar heat for the apartment. The cost of the solar collection and storage is thus part of the cost of building and main-

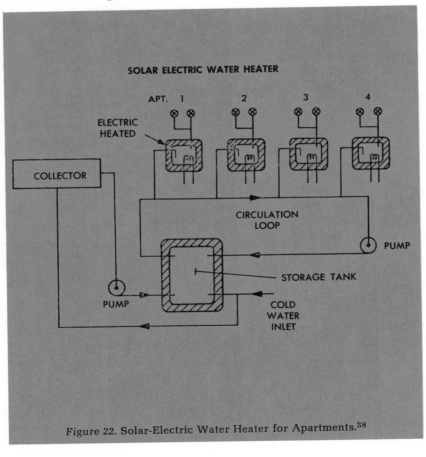

Figure 22. Solar-Electric Water Heater for Apartments.[38]

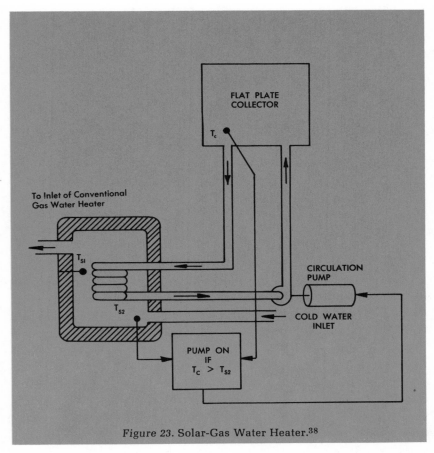

Figure 23. Solar-Gas Water Heater.[38]

taining the apartment building, and is included in the rent. The electric power consumed, however, is paid for by the individual user in his electric bill. This aspect of the system is attractive from the viewpoint of the apartment owner since it provides accountability for the consumption of hot water during periods when the solar input alone is not adequate. The same general type of solar collector can be used to preheat water before it enters a conventional gas water heater. Water heating in a freezing climate can utilize an intermediate heat transfer fluid (antifreeze solution) circulating through the collector in a closed loop to transfer its heat to water in a heat exchanger (Figure 23). If the collector temperature is higher than the cold water inlet temperature, which is usually the case when the sun is shining on the collector, the pump is turned on and fluid from the collector circulates through a coil in the storage tank, thus preheating the water in the tank before it enters the conventional heaters. Solar heat is thereby used year round to reduce the consumption of gas or electricity for heating water. During parts of the summer all of the heat can be supplied by the solar collector. For the closed loop system the water flowing through the collector can be maintained at a lower pressure than the water in the storage tank.

It is clear that as gas prices rise, and as solar collector costs decrease, solar-assisted gas water heaters will become cheaper than gas heaters alone. Already, solar-assisted electric heating is cheaper than electric water heating alone because of its high cost. At the present time gas is not being supplied to new units, because of short supply, in some areas of the country.

Solar water heating has been quite popular in Israel, and by 1965 over 100,000 units had been installed.[48] The first solar water heaters in Israel were sold with a three-year guarantee. This was soon raised to 5 years, and for a small additional cost could be extended to 8 years.

Figure 24 shows a typical natural convection water heater. A small reverse flow occurs when the collector is cooler than the water in the tank, which has a certain advantage in the winter because it prevents water from freezing in the collector. One series of tests measured the overall performance as affected by seasonal variations, type of transparent covering, insulation, height of storage tank and location of the point joining the flow pipe to the storage tank. The average efficiency was about 50% with a polyvinyl fluoride collector covering, and about 55% with glass. There was little effect on the efficiency of changes in insulation and seasonal variations.[49]

In the United States solar water heating should probably be utilized in all new apartment and housing units in areas with mild winters. In more northern climates where winter temperatures drop well below freezing, natural circulation systems such as shown in Figure 24 cannot be used.

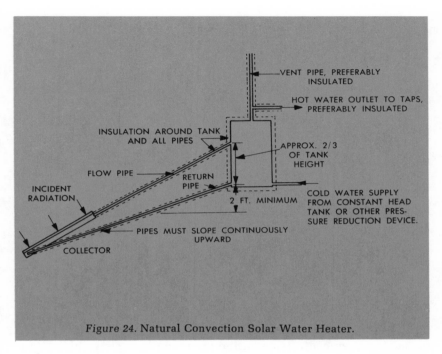

Figure 24. Natural Convection Solar Water Heater.

There are now a number of U.S. manufacturers of complete packaged solar water heating systems, including Revere Copper and Brass, Grumman Aerospace, Sunworks, and others. These companies have local distributors in most parts of the country who can supply the latest information and prices for their units.

Chapter 6

AIR CONDITIONING

Solar cooling is usually accomplished by using solar heat to operate a thermal absorption type refrigeration system. In regular electrically operated refrigerators, a vapor such as ammonia is condensed to a liquid with a motor-driven pump, and the heat released is removed by blowing air through the condenser. The liquid is then vaporized in an insulated box and heat is absorbed by the vaporization process to give the cooling effect. In solar refrigeration, the ammonia cycle is similar except that the pressure is produced by heating a concentrated solution of ammonia to give a high vapor pressure, instead of compressing the vapor mechanically. There are two connecting, gastight vessels, one of which contains liquid ammonia and the other a very concentrated solution of salt in liquid ammonia. The salt solution has a much lower vapor pressure. The liquid ammonia vaporizes in its compartment, thereby cooling it, and dissolves in the salt solution contained in the other compartment. The system is regenerated by using solar heat to raise the temperature of the salt solution so the vapor pressure of ammonia in the solution exceeds the vapor pressure of the pure liquid ammonia in the second compartment. In this way, the operating cycle produces cooling by evaporating ammonia as it goes into the concentrated solution of salt, making it more dilute; and the solar regeneration drives out the ammonia from the diluted salt solution to produce pure ammonia and leaves a more concentrated solution. Another cycle uses a concentrated solution of lithium bromide to absorb water vapor which causes liquid water in another compartment to vaporize to provide cooling. The lithium bromide solution is concentrated again by heating the dilute solution with solar heat, and the system is operated continuously.[50]

A continuously operating absorption air-conditioning system was built and tested in the early 1960's at the University of Florida.[51] Hot water was used to heat a high-concentration, ammonia-water solution (50%–60% ammonia by weight) in a generator, driving the ammonia out of the solution. The ammonia vapor was then condensed and expanded through an adjustable expansion valve, entering the evaporator as a two-phase mixture. The liquid component evaporated, cooling the water circulating through tubes in the evaporator, and then was reabsorbed into the water, and the ammonia solution was pumped back to the generator to repeat the cycle. Ten 4 ft × 10 ft flat plate solar collectors provided the hot water to operate the air conditioner. The absorbing surfaces were tubed copper sheets painted flat black, placed in galvanized sheet-metal boxes with two inches of foam-glass insulation behind, and a single glass cover. The system was operated with heating water temperatures ranging from 140–212°F. The maximum cooling effect was 3.7 tons, and steady operation was achieved with 2.4 tons of cooling.

Solar-powered air conditioning systems can also be driven by an organic Rankine cycle engine. Solar heat could be used to vaporize an organic fluid at a temperature between 160°F and 280°F to drive a Rankine cycle engine, which in turn would drive the compressor of a vapor-compression air conditioning system.[52] The coefficient of performance should compare favorably with absorption air conditioning systems, but at the present time none are commercially available. Except for the northern-most part of the country, combined solar heating and cooling is usually cheaper than solar heating or cooling alone. Solar cooling permits both summer and winter utilization of the solar collector—the most expensive part of the system. Thus, in most parts of the U.S., the economics favor combined solar heating and cooling, rather than either alone. Figure 25 illustrates such a combined system using a common collector, storage tank, auxiliary heater, and blower for both heating and air conditioning.

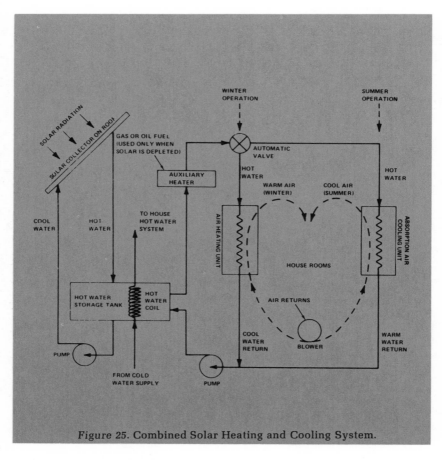

Figure 25. Combined Solar Heating and Cooling System.

Water-fired chillers suitable for solar air conditioning systems are currently available from York and Carrier for applications requiring 100 tons or more of cooling, and units rated at 3 tons and 25 tons are available from Arkla-Servel. The 3-ton Arkla-Servel SOLAIRE-36 was listed at $2750 in 1976.

Chapter 7

ELECTRIC POWER GENERATION

A variety of approaches have been used for converting solar energy into electricity, including solar-thermal conversion, photovoltaic devices, and bio-conversion. Sunlight is an abundant, clean source of power; all that is required is the development of technology to economically convert this energy into electricity.

The NSF/NASA Solar Energy Panel[13] identified the various possible steps leading from solar radiation to power delivered to the consumer (Figure 26). In this scheme plants, rivers, winds, ocean currents and ocean temperature gradiants are considered natural collectors of solar energy. Solar energy can also be collected directly as heat, or converted into electricity via the photoelectric effect. If collected as heat, it can be stored for use when the sun is not shining. The heat can be used to operate a power plant or to produce a chemical fuel, such as through the thermochemical production of hydrogen. The fuel can be stored and used as needed to produce electric power, such as with a hydrogen-air fuel cell.

With so many possible approaches available for the production of electric power, the problem then is to choose the approach which is most cost-effective for a specific application. This is sometimes difficult since technology is advancing rapidly in most of these areas, and the comparative economics becomes uncertain. At present, the two most promising technological approaches are photovoltaic conversion with electrical storage, and solar-thermal conversion with heat storage for night-time operation.

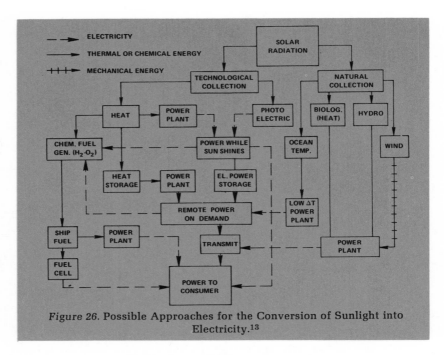

Figure 26. Possible Approaches for the Conversion of Sunlight into
Electricity.[13]

Solar-Thermal Power Generation

The two main approaches to solar-thermal power generation are
the solar furnace approach, in which sunlight reflected from many
different locations is concentrated on a single heat exchanger, and
the solar farm, with large numbers of linear reflectors focusing solar
radiation on long pipes which collect the heat.

The tower concept proposed in 1949 is a good example of the
solar furnace approach.[53] A large number of flat mirrors covering a
large area of land independently focus sunlight onto a boiler, which
is mounted at the top of a tower located near the center of the field
of mirrors to produce high-temperature steam for driving a turbine.
A 100-kilowatt plant has been built and operated in Italy,[54] and a
400-kilowatt plant of a similar design is being built at the Georgia
Institute of Technology. An advantage of this system is that separate mirrors and steering mechanisms can be mass produced, and
the smaller reflectors are less subject to high wind loadings than

a single large steerable concentrator of the same total collector area.

A system has been proposed with over a thousand 10-ft mirrors covering a 6000-ft diameter circle of about one square mile area to reflect sunlight onto the boiler on top of a high tower (Figure 27). Each mirror would be steered separately by a heliostat as shown

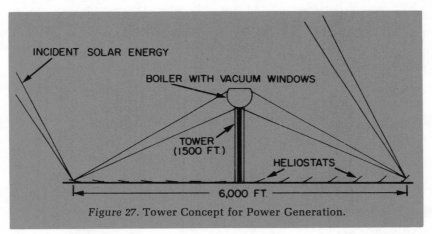

Figure 27. Tower Concept for Power Generation.

in Figure 28. Since the major expense of solar energy collection employing a solar furnace would be the heliostats, considerable research is being done in order to develop a heliostat which can be economically mass produced. The boiler could be made of steel and operate in the 1000°C range, and the solar image size at the boiler would be 31 ft in diameter. If 45% of the land area is covered with mirrors, the boiler could collect 630 BTU/ft²/day of mirrors in the Southwest U.S. in the winter, 1320 BTU/ft²/day in the spring and fall, and 1620 BTU/ft²/day in the summer.[55] The heliostats must automatically aim the mirror with an accuracy of 0.2° in the presence of winds.

A megawatt solar furnace was built in France employing heliostats with 20-in. square flat glass mirrors and a fixed parabolic concentrator on the side of a nine-story building.[56] The flat mirrors reflect sunlight toward the fixed parabolic concentrator that focuses the sunlight. The heliostats and mirrors cost $21/ft².

Another similar power plant system uses arrays of heliostat-guided mirrors to focus sunlight into a cavity-type boiler near the

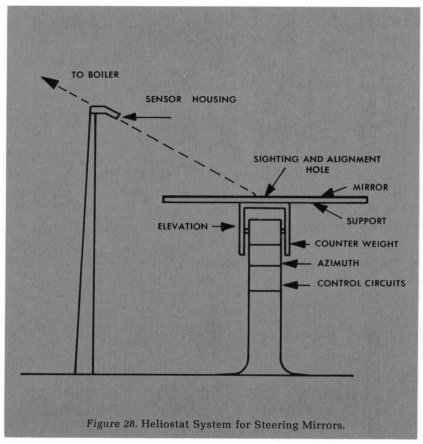

Figure 28. Heliostat System for Steering Mirrors.

ground to produce steam for a steam turbine electric power plant. Sunlight striking the mirrored faces of the heliostat modules is reflected and concentrated in the cavity of the heat exchanger. To demonstrate the feasibility of generating large amounts of electric power with this system, Martin Marietta and the Georgia Institute of Technology, with support from ERDA are generating about 300 kilowatts of power using the French furnace in Odellio.[57] One useful application for this system would be to augment existing hydroelectric plants whose generating capacity exceeds their water supply. With solar augmentation the power production for the central six hours of daylight is taken over by the solar system, so hydroelectric generation during the remaining hours is augmented

33% by the water saved during solar operation. The solar plant does not need to be anywhere near the hydroelectric facilities as long as it is connected to the same power grid. This solar plant can also be used for peaking in areas where peak power demand is during daylight hours.

Solar farms have been proposed using parabolic trough concentrators or other types of concentrators to focus sunlight onto a central pipe surrounded by an evacuated quartz envelope. Heat collected by a fluid flowing through the pipes could be stored at temperatures over 1000°F in a molten eutectic and used as required to produce high-enthalpy steam for electric power generation.[58] Another approach is to store the heat in rocks, and extract the heat as required to generate steam on demand, as illustrated by Figure 29.

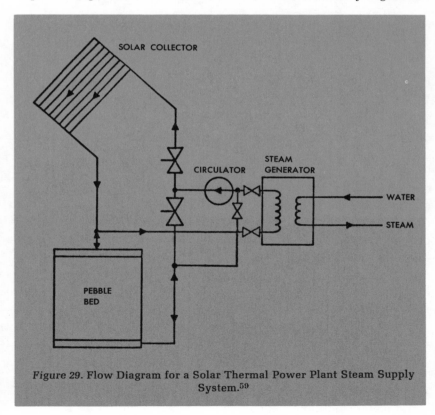

Figure 29. Flow Diagram for a Solar Thermal Power Plant Steam Supply System.[59]

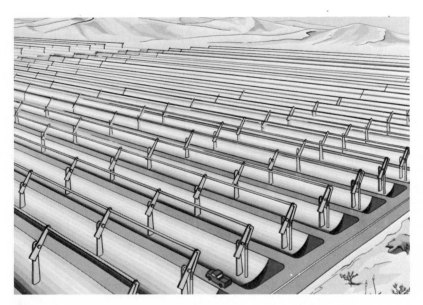

Figure 30. Artist's Concept of Fixed-Mirror Solar Concentrators Showing the Mirrors and the Tracking Heat Absorber Pipes.[59]

The fixed mirror concentrator does not have to be steered and need not be self-supporting, so fabrication of these concentrators should be cheaper than for steerable reflectors.[59,60] Since the point of focus always lies on the reference cylindrical surface, the heat exchanger pipe can be supported on arms that pivot at the center of the reference cylinder. This greatly simplifies the positioning of the heat exchanger. Figure 30 illustrates the fixed-mirror concentrator array. Costs of power from this facility are estimated to be competitive with alternative means of power generation.[59] Land costs would be negligible since, even at $1000/acre, the land cost is only $0.023/ft^2. Desert land is even cheaper.

A 1,000,000-MWe solar-thermal power plant has been proposed that would cover about 13,000 square miles of desert extending from the upper regions of the Gulf of California as far north as Nevada (Figure 31). The plant would use waste heat to produce 50 billion gallons of water each day, enough to meet the needs of 120,000,000 people. The proposed plant would use a circulating liquid metal (sodium or NaK) to extract heat from a solar farm and

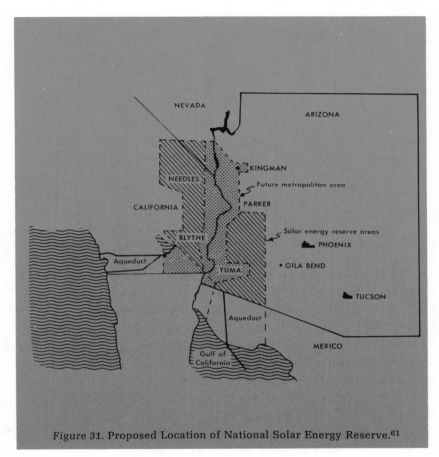

Figure 31. Proposed Location of National Solar Energy Reserve.[61]

store it in a phase-change salt or eutectic mixture at temperatures in excess of 1000°F. Power would be produced by a high-pressure steam turbine-generator, and the low-pressure steam from the turbine used to distill water. The total cost of solar heat collected by this plant is estimated at $0.50 per million BTU.[61]

Photovoltaic Power Generation

Solar cells offer a potentially attractive means for the direct conversion of sunlight into electricity with high reliability and low maintenance, as compared with solar-thermal systems. The present

disadvantages are the high cost of about $25/watt,[62] and the diffi-
culty of storing large amounts of electricity for later use as compared with the relative ease of storing heat for later use. The cost
of solar cells is expected to be considerably reduced when cells are
manufactured in large quantities using new production techniques
for obtaining ribbons or sheets of single crystal silicon. At present
large crystals of silicon or other semiconducting material are grown
and then sliced into thin cells; new techniques for producing the
thin slices directly use edge-defined film growth,[63] dendritic
growth,[64] rolled silicon,[65] or sheets of cast silicon that are recrystal-
lized through heated or molten zones.[66] Silicon itself is very cheap
since it is the second most abundant element in the earth's crust,
and is produced in the U.S. at an annual rate of 66,000 tons at a cost
of $600/ton. Therefore, when the most suitable of these mass man-
ufacturing techniques is utilized, the cost of solar cell arrays
should be reduced to $1/watt or less, making them useful for the
large scale generation of electric power.[63, 67] Cadmium sulfide solar
cells also offer great potential for low-cost manufacture.

Four companies which manufacture solar cells are Heliotech,
Centralab, Solar Power Corporation (Exxon), and Sharp. Solar
Power Corporation[68] sells a small solar power module that produces 1.5 watts at a solar intensity of 100 mW/cm². The current and
power output characteristics of these solar cells (typical of solar
cells in general) are given by Figure 32. Standard conditions are
0°C and 1000 W/m² insolation, typical conditions are 25°C and 800
W/m² insolation. The solar array module consists of five 2.17-in.
diameter silicon solar cells attached to a 13.5 in. × 9 in. panel and
is usually used to charge storage batteries to provide a continuous
supply of power in remote locations. Tests in Arizona showed no
degradation in output over a six month period. One power system
being used at present to power navigational lights consists of 80 of
these modules, 28 100-amp-hr 12-volt storage batteries, and the
electronic control circuit. This power supply is cheaper to use than
the alternatives; the Coast Guard saves about $3 million/year by
using solar-powered buoys.[69] The cost reduction is mainly due to
the smaller number of trips out to the buoys for servicing. Wires
are used to prevent seagulls from landing, but nothing is done about

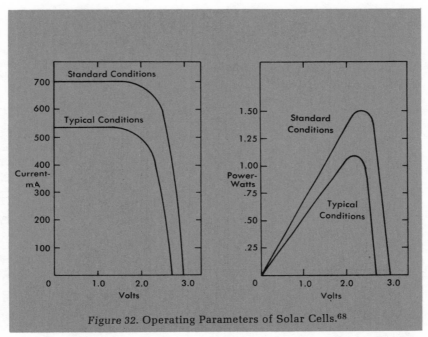

Figure 32. Operating Parameters of Solar Cells.[68]

snow. NASA's experience testing solar cell arrays in Cleveland has shown no significant reduction in power due to dirt or dust accumulation and little problem with snow.

Large solar cell arrays have been considered for supplying the electric power needs of the western United States in 1990, assuming that solar cells can be mass produced at $1/watt. An array covering 192 square miles, coupled with pumped storage, would supply the 14,300 MWe needed by Arizona in 1990 for about $58 billion, and an array covering 2200 square miles (44 × 50 miles) would supply 40% of the electrical power needs of the 11 western states for a capital cost of around $637 billion.[70] Since these costs are in excess of alternative means of power generation, it appears that even at $1/watt solar cells may be too expensive for central station power generation. The cost of solar cells should be reduced to about $0.25 per watt before solar cell arrays can be expected to become practical for central station power generation.[71]

The cost of generating electric power with solar cells can be reduced by using concentrators to focus sunlight onto the cell. One simple type of concentrator is the reflecting cone (Figure 33). With-

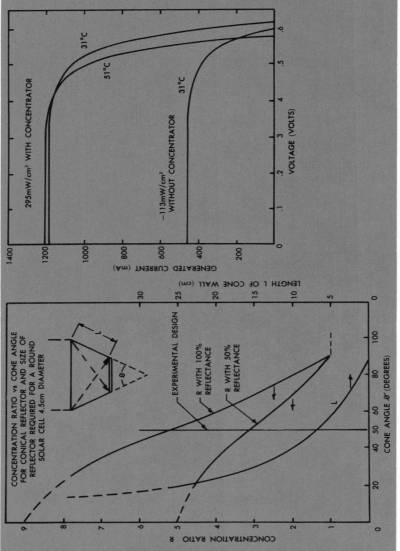

Figure 33. Concentration Ratio of Cone Reflector and Effect on Solar Cell Performance.[72]

out external cooling concentration ratios of up to five can be used without seriously reducing the cell performance due to cell heating. Higher concentration ratios are possible with external cooling. Solar cell arrays with concentrators must be steered to follow the sun; in the case of the conical concentrator the output is reduced below that with no concentrator if the angle of incidence is less than 60°, and at angles of less than 45° the output is negligible.[72]

Another related design is the channel concentrator consisting of two flat reflecting surfaces at an angle of 30° placed on both sides of a line of solar cells. The theoretical maximum concentration ratio is 3; an actual concentration ratio of 2.25 has been achieved with a channel concentrator array using 2 in. × 2 in. silicon solar cells at the base of the V channel.[62] Five channels with 30 cells each formed a 4.75 lb, 1 ft × 2 ft array producing 12 watts at 12 volts.

With external cooling, silicon solar cell outputs can be increased by more than 100 with concentrating systems.[73] Using experimental data[74] for cells operating with solar fluxes between 14 and 28 watts/cm^2, a system was designed to produce 50 watts of electric power from 36 square centimeters of cell area by using a 5½-ft parabolic concentrator to provide a solar flux of 28 watts/cm^2.[75] The cells were water-cooled to maintain their temperature at 200°F. Five watts were required to pump the water. A 250-watt electric power plant in the Soviet Union used a concentrator consisting of 26 plane mirror facets forming an approximate parabolic cylinder.[33] The concentrator increased the power output a factor of 5.2 over the power output with no concentrator, the solar cells were water-cooled, and the overall plant efficiency was 2.7%.[76] Another plant was developed by the same group using channel concentrators with a concentration ratio of 2.5, and not requiring water cooling. These plants were developed to provide power for water pumps in the grazing areas of the southern regions of the U.S.S.R. One of them was installed at the Bakharden state livestock-breeding farm situated in the Kara-Kum Desert, Turkmenia. Its output equals about 400 watts, enough to lift from a depth of 20 meters a sufficient amount of water for 2,000 sheep.[77]

Chapter 8

TOTAL ENERGY SYSTEMS

The feasibility of using solar energy to provide for all of the various energy needs of a home, business, or community requires either the development of inexpensive solar cells or an economical means of collecting solar heat at high temperatures and converting it to electric power. Photovoltaic cells can be combined with a flat plate collector so that the radiant energy not converted into electric power is collected as heat and used to supply hot water, space heating, absorption refrigeration, and air conditioning. Figure 34 illustrates a solar cell flat plate collector which would permit utilization of up to 60% of the available solar energy. Collectors mounted on vertical walls and/or part of the roof of a house or apartment building can supply all the various types of energy needs of the building. Figure 35 is a schematic showing the energy flows for a residential total energy system using solar cell flat plate collectors. This type of system is perhaps the ultimate in residential solar energy utilization, since both heat and electric power are produced without any moving parts, except for the pump or blower circulating coolant through the collector.

Advantages of this type of solar electric-thermal total energy system are: 1) The collector uses the same land area as occupied by the building, and thus there is minimal effect on the environment through use of land presently being used for other purposes, 2) About three times the present average household consumption of electric power can be collected from average-size family residences, even in the northeastern U.S. (This surplus energy could be used for charging an electric automobile), 3) The system is not vulnerable

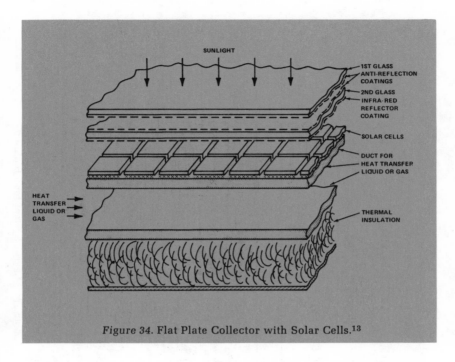

Figure 34. Flat Plate Collector with Solar Cells.[13]

to breakdowns in central energy generation stations or transmission systems, and 4) The small size of the individual unit makes prototype testing and demonstration relatively inexpensive, and will help to attract consumer-oriented industries.

Researchers at the University of Florida have built and tested solar water heaters, solar air heaters, a solar still, a five-ton solar air conditioner, a solar refrigerator, several solar ovens, a solar sewage digester, solar cell arrays, several types of solar powered hot air engines, solar water pumps, a "solar-electric" car, and a solar house.[78] The solar house, occupied by a graduate student and his wife, uses solar energy for space heating, water heating, swimming pool heating, electricity, and recycling of liquid wastes with the solar still. A ⅓-horsepower hot air engine operating from a 5-ft parabolic concentrator drives a dc generator to charge the solar-electric automobile to provide pollutionless transportation from the solar house.[79] Thus it has been shown that it is technologically possible to use solar energy to provide all residential energy needs.

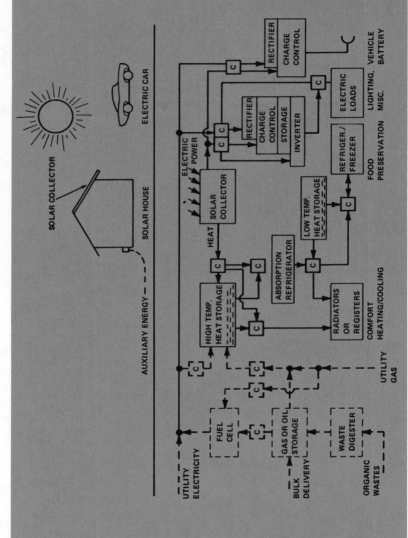

Figure 35. Schematic of a Solar Total Energy System for a Building.[13]

If inexpensive solar cells are manufactured, then the major remaining obstacle to the development of total energy systems is the problem of storing the electricity. One approach is to store the electricity in the form of hydrogen. Excess electric power generated during the day is used to electrolyze water to produce hydrogen and oxygen gas, which is compressed into storage tanks, and used in the evening with a hydrogen-oxygen fuel cell. This system is attractive in the long run, but too expensive at present for residential use.[80] Another possible energy storage medium is the flywheel. A new safe flywheel has been proposed with an energy storage capacity of 30 watt-hours per pound.[81] Excess electric power generated during the day is used to increase the rotational velocity of the flywheel, and in the evening the energy of the flywheel is used to generate electric power. Lead-acid batteries could be used, but if they were used to store a substantial fraction of all the electrical energy produced in the United States, it is questionable whether enough lead would be available.[82] Other electrochemical systems, however, might be possible.

The other approach to developing a total energy system, not involving solar cells, is to collect the heat at a high temperature using a dynamic conversion system to produce electric power, and use the waste heat for space heating and cooling. One group studied four different types of total energy systems using concentrating collectors, high-temperature heat storage, and a derated turbine, where the exhaust energy is used for heating and air conditioning. Another system with a flat plate collector driving an organic turbine generator was rejected as not being economically competitive with focused concentrator systems. They calculated the performance and economics of each proposed system for Albuquerque, N.M.[83-85] One day in four was assumed cloudy and the direct insolation taken to be 80% of the total. The cost of the residential solar energy systems were compared with a "normal" system supplying equal energy demands with utility electric power, and natural gas for space heating, air conditioning, and water heating. The results of these calculations indicate that solar total energy plants with high temperature collection and three levels of heat storage would be economically competitive with the "normal" system when the wholesale fuel cost reaches $0.90/million BTU.

Large users of energy such as apartment complexes, shopping centers, and industries can take advantage of solar-thermal total energy plants ranging in size from 0.2 to 20 megawatts. As of 1972 there were about 550 total energy plants in this size range in operation in the United States.[86] The more recently installed plants have averaged over five megawatts in capacity. Electrical storage problems for all types of total energy plants can be reduced considerably if the electric utility company owns and maintains these systems, and allows the excess power generated during the day to be fed back into the utility power grid. The electric power company could then give credit on the electric bill for power supplied by the customer.

Chapter 9

INDUSTRIAL AND
AGRICULTURAL APPLICATIONS

The parabolic concentrator has provided a means of generating very high temperatures for small-scale industrial applications and for research purposes, and solar heat at lower temperatures has been used for both industrial and agricultural drying operations. These represent two of the more promising commercial uses of solar energy.

Solar Furnaces

A megawatt furnace was built in France in the 1950's using heliostats to direct sunlight toward a large parabolic concentrator[56] (Figures 36 and 37). A similar 70-kilowatt furnace was built in Japan using a 10-meter-diameter parabolic concentrator. Another furnace of the heliostat type in Nantick, Mass., uses a spherical concentrator.

The Japanese furnace began operation in 1963 and produces temperatures in excess of 3400°C, the melting temperature of tungsten. Refractory bricks have been melted even in feeble sunlight. The furnace is used for studies of high-temperature materials properties and some manufacturing. For example, alumina when melted in a graphite cylinder assumes a spherical shape because of its large surface tension. Turning the cylinder properly results in the formation of a fused alumina crucible which has much more desirable properties than a sintered one. Tungsten melted in an inert gas does not form a carbide even though the melting occurs on a graphite surface. Front surface aluminized mirrors used for the furnace showed a reduction in reflectivity from about 95% to 85% over a

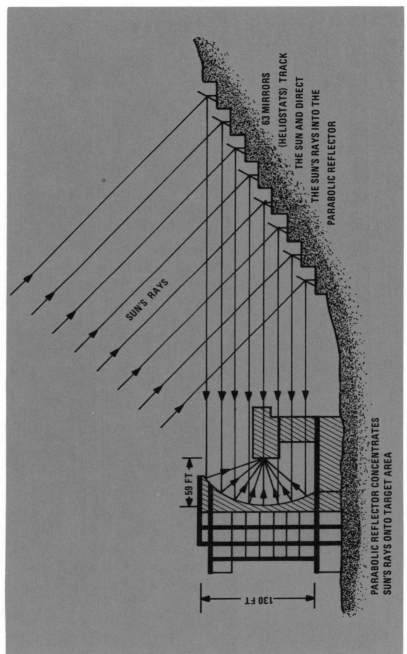

Figure 36. Schematic of 1000-kw Solar Furnace.

Figure 37. French Solar Furnace. (Courtesy of J. D. Walton)

five-year period. At the present time all the mirrors are aluminized once each year.[87]

The French solar furnace is located in the Pyrenees at Odeillo-Font Romeu (altitude, 5900 feet), about 20 miles east of Andorra. At this location the sun shines as many as 180 days a year and solar intensities as high as 1000 watts per square meter are common. The solar furnace was completed in 1970 at a cost of about $2,000,000.

The parabolic reflector has a focal length of 59 feet, is 130 feet high and 175 feet wide and is composed of 9500 mirrors 17.7 in. × 17.7 in. Since the parabolic reflector is too large to track the sun, 63 heliostats set in eight tiers are used to follow the sun and reflect its ray in parallel beams onto the parabola. The heliostats are 24.6 × 19.7 ft. and each is composed of 180 mirrors 19.7 in. × 19.7 in. by 19.7 inches.

The solar energy incident on an area of about 23,000 square feet is concentrated by the parabolic reflector into an area about two feet in diameter. Sixty percent of the total thermal energy (about 600 kilowatts) is concentrated in an area one foot in diameter at the center of the focal plane of the parabola.

A similar but smaller solar furnace in the Soviet Union melts refractory materials at a temperature of up to 3,500°C, and is used for producing high-purity refractories.[77]

Air Heaters

Solar air heaters have a great potential for improving agricultural drying operations around the world. At present a large portion of the world's supply of dried fruits and vegetables continues to be sun-dried in the open under primitive conditions. Being unprotected from unexpected rains, windborne dirt and dust, and from infestation by insects, rodents and other aninmals, the quality is often seriously degraded, sometimes beyond edibility. In an increasingly hungering world, practical ways of cheaply and sanitarily preserving foods are needed.

There are two basic methods of solar dehydration. By the first method the necessary heat is supplied by directly exposing the material to solar radiation, which also enhances the proper color development of greenish fruits by allowing, during dehydration, the decomposition of residual chlorophyll in the tissue under direct solar radiation. The major drawbacks are the posible damage due to overheating, and relatively slow drying rates resulting from poor vapor removal in cabinet driers. According to the second method the foodstuff is heated by circulating preheated air. Since the drying material is not subjected to direct sunshine, caramelization and heat damage do not occur. A further advantage is that the circulating air absorbs the water vapor from the food, thus accelerating drying. On the other hand, products of inferior appearance may result if immature fruit is dehydrated, since shading prevents the breaking down of chlorophyll.

One dryer used a square meter area of steel chips beneath a glass cover to absorb solar radiation, and passed air to be heated through the chips. Steel chips are cheap, have a high heat transfer area per unit volume and excellent turbulence geometries, and an absorptivity of 0.97. Several agricultural products were dried and compared with an open-air sun-dried control group. Peppers dried in the solar dryer had attactive bright colors as opposed to the brownish color of the slower drying control batch, which was sun-dried in the open. Similarly, for the dehydration of sultana seedless grapes, the sun-dried control sample was rained upon, causing it to have a dark color. Soon afterward it was attacked by birds so the weighings were terminated. Raisins in the deydrator had a golden color and were dried in spite of continuous rainy weather.[88]

A variety of solar heaters have been developed for use in crop drying, space heating, and for regenerating dehumidifying agents. These various types of heaters provide air at 100°F above ambient with collection efficiencies of 50% or more. The heat transfer processes in air heaters are quite different from those in flat plate collectors which heat water. In the water-cooled collector, heat absorbed by the plate is transferred to the water tubes by conduction, so the absorber plate must have a high thermal conductivity. In an air heater the air can be in contact with the whole absorbing surface, so the thermal conductivity of the absorbing surface is of little importance. This makes solar collectors for heating air inherently cheaper than solar collectors for heating water. The main factors determining the efficiency of heat collection of a solar air heater operating at a given air inlet temperature are:

1. Heater configuration, that is, the aspect ratio of the duct and the length of duct through which the air passes.
2. Air mass flow rate through heater.
3. Spectral reflectance and spectral transmittance properties of the absorber cover.
4. Spectral reflectance properties of the absorber plate.
5. Stagnant air and natural-convection barriers between the absorber plate and ambient air.
6. Heat transfer coefficient between the absorber plate and the air stream.
7. Insulation at the absorber base.
8. Insolation.

V-corrugation of the absorber plate considerably improves the performance over that of collectors with flat absorbing surfaces. Spectrally selective coatings also improve performance. Air heaters of simple construction employing cheap materials have been shown to be capable of supplying air at temperatures above 150°F with good efficiency.[40] For crop drying only air temperatures below 180°F are needed.

One study of flat plate air heaters with two glass covers showed that if the air is passed between the two glass panes before passing through the blackened metal collector the outer glass temperature is reduced 4°F to 10°F, the collection efficiency increases 10% to 15%, and the temperature rise of the air is increased as much as 20%. Thus, it appears an attractive nonconcentrating air heater design could use the two pass configuration and a V-corrugated absorber with spectrally selective coating.

Tests were made of air heaters with an absorber consisting of 96 parallel specularly reflecting aluminum fins 6.35 cm high, 0.635 cm apart, and 61 cm long. A single 0.317-cm glass coverplate was placed over the absorber, and air pumped between the fins. The collectors measured 61 cm × 61 cm. The collector with specularly reflecting fins was shown to be about 15% more efficient than an identical collector with diffuse fins.[89] Solar air heaters using hot water from water-cooled flat plate collectors have also been built.[90]

The use of concentrators to produce higher air temperatures for industrial operations, such as the 250° to 500°F needed by textile mills, has received little attention so far. Parabolic cylinder concentrators or faceted concentrators such as proposed by Russell could be used to heat air to high temperatures. The hot air can be used directly for textile or agricultural drying and other industrial operations requiring hot air at temperatures up to 1000°F.

Chapter 10

SOLAR STILLS

Solar stills are being used increasingly worldwide to produce drinking water from salty or polluted water. A still at the University of Florida is used to reclaim drinking water from household liquid wastes.[78] Solar stills are the cheapest means for desalting quantities of less than 50,000 gal of saline water per day in areas of reasonable sunshine, and production costs are currently about $3.50/1000 gal.[91]

A solar still is typically a transparent plastic tent or glass enclosure containing a shallow pan of saline water with a black bottom. Sunlight heats the water in the pan, causing it to evaporate and recondense on the underside of the sloping plastic or glass and run down into collecting troughs along the inside lower edges of the transparent cover. The performance of solar stills has been calculated under various conditions of ambient temperature and insolation, and the results showed close agreement with data from a 4500-ft² solar still located at Muresk in Western Australia (Figure 38).[92] The daily output rose from about 0.1 lb/ft² (450 lb total) of water per day in the winter (July) to about 0.8 lb/ft² (3600 lb total) of water per day in the summer (December), so the range of production for this still is from 0.012 gal to 0.1 gal/day/ft² of collector. Similarly, a large 23,300-ft² solar still[93] on the island of Saint Vincent in the West Indies provides the most economical source of fresh water (other than rainwater), since underground natural sources are not available and the cost of shipping water to the island is high. The average daily output of the plant is about 0.05 gal/ft² of collector, or more than 1000 gal/day for the plant. Four-mil polyvinyl fluoride film is used as the transparent cover.

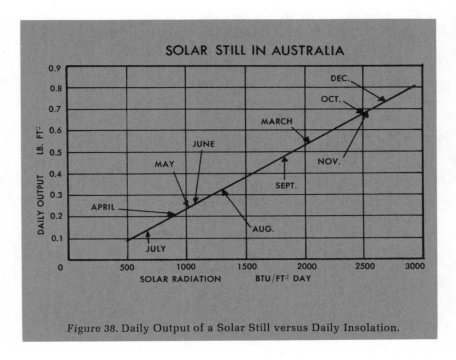

Figure 38. Daily Output of a Solar Still versus Daily Insolation.

For buildings solar stills may be mounted on rooftops (Figure 39).[94] An advantage of this approach is that the cost of the solar

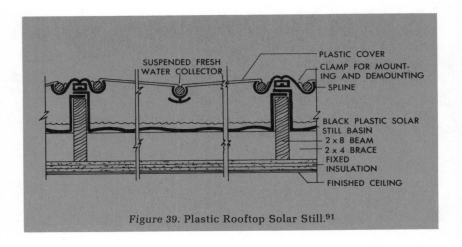

Figure 39. Plastic Rooftop Solar Still.[91]

still is partially offset by the savings in ordinary roof costs, since the still replaces the roof. Also, the still is not occupying land area that could be put to other uses. Since solar collectors for space heating and cooling require only about half the roof area, the rest of the roof could be a still to provide fresh water. Shallow depths of saline water are used for maximum yield, so unlike the roof pond, the solar still would add little to the weight of the roof. The stills can use 4-mil polyvinyl fluoride film, treated on the underside for wettability, which should have at least a 5-year life. The inherent safety hazards of glass restrict its extensive use on roofs in densely populated areas, and breakage could cause puncture of the water-tight liner with subsequent flooding of the room beneath. With proper design, replacement of the PVF cover every five years should be a simple matter. Plastic pipes and fittings would be used to reduce cost and weight.[94]

The still in Figure 39 uses a rigid basin of molded resin supported directly on the 2 × 4 in. braces between the ceiling beams. The PVF transparent cover film is fastened with S-clamps onto the main 2 × 8 in. roof beams. The weight of the center-suspended condensate collector contributes to the vapor seal and shapes the V-cover so that distillate drains to the collector. The result is an inexpensive waterproof roof which provides a supply of fresh water. Accidental cover damage would, at worst, allow rain to drain into the condensate collector. PVF covers have produced the highest yield for solar stills.[95]

In the Soviet Union large solar stills are used both for industrial and agricultural purposes. A reinforced concrete still was built in 1970 in the Shafrikan collective farm at Bukhere Oblast in the Uzbek Republic.[96] The water in that area was unusable for many purposes because of its very high mineral and sulfur content. With an evaporative area of 6500 ft^2, the still produces a yearly average of 0.08 gallons/ft^2/day, a total of about 540 gal/day·on the average. The still consists of 39 glass-covered independent sections of 168 ft^2 each with a trough depth of 10 cm. The maximum output of 80 gal/hour occurs between 2 PM and 4 PM (in August) and a minimum output of 5 gallons/hour is produced between 3 AM and 7 AM. Another large still uses steps inclined at a 2°–3° angle so the water flows over the steps, from upper to lower, until it reaches the discharge drain. This flow enhances evaporation and increases the output and efficiency of the still about 20%.[97]

The Krzhizhanovsky Power Institute in Moscow has also been studying various aspects of solar stills. Theoretical studies were conducted of heat and mass transfer processes in solar stills of the hotbox type and techniques for calculating the performance of these stills were developed. In a properly designed still most of the solar energy that passes through the glass (or film) is used to evaporate saline water. As a result, the space within the still is filled with a steam-air mixture. The energy balance conditions during operation of the still are such that the surface of the glass is at a lower temperature than that of the steam-air mixture, with the result that water vapor condenses on the glass surface, whereas the condensate runs down the inclined glass, drips into the groove and is collected in the tank. Well-instrumented solar stills were constructed to investigate these processes. During the tests the temperature of the water heated by the sun varied from 74°F–207°F while the temperature of the glass condensing surface varied from 61°F–192°F. As a result of these studies equations were developed which accurately describe heat and mass transfer processes in this type of solar still.[98]

The effect of wind speed and direction on the output of a solar still of the greenhouse (glass) type was studied by using a fan to blow air across a small still. For saline water temperatures of 104°F, 131°F and 158°F the wind speed was varied from 0–26 ft/sec at wind directions of 0, 45, 90, 135 and 180°; and for all wind directions and temperatures the maximum still output was achieved for a wind velocity of about 16 ft/sec. Increasing the wind velocity up to this value increases the rate of heat removal from the glass cover, which increases the rate of condensation on the glass, resulting in an acceleration of the evaporation process and as much as a 25% increase in still output. Further increases in wind speed lead to a reduction in the saline water temperature which reduces the evaporation rate and still output. The most favorable wind direction is parallel to the condensing surfaces (90° angle).[99]

Chapter 11

CLEAN RENEWABLE FUELS

Most of the energy used in the United States today comes from fossil fuels produced many years ago from solar energy. Clean renewable fuels to supplement and eventually replace these fossil fuels can be produced from plant life grown under more optimum conditions than found in nature, and from organic waste materials. The various processes for the production of these fuels listed in Figure 40 are aimed at converting organic materials with a low heating value per unit weight into higher heating value fuels similar to the fossil fuels currently in use. Another possible technique is the use of high-temperature heat from solar concentrators to operate a regenerative thermochemical cycle for the production of hydrogen; the hydrogen can be used directly or utilized for the production of hydrocarbon fuels such as methane.

Perhaps the oldest and simplest technique for the production of a clean renewable fuel is to grow plants and burn the plants for energy; this could be done on a large scale for electric power generation.[100] Air pollution from such a plant is minimal since virtually no oxides of sulfur are produced, particulate emissions can be controlled with precipitators, and the CO_2 released is reabsorbed by the growth of new plants. Up to 3% of the incident solar energy can be absorbed by plants, and this energy is released when the plants are burned.[101] For a 1000-MWe steam-electric power plant operated at a load factor of 75% with a thermal efficiency of 35%, 150 square miles of land area is required to fuel the plant if the average insolation is 1400 BTU/ft²/day and the capture efficiency of the plants is 3%. It has been calculated that the total cost of the

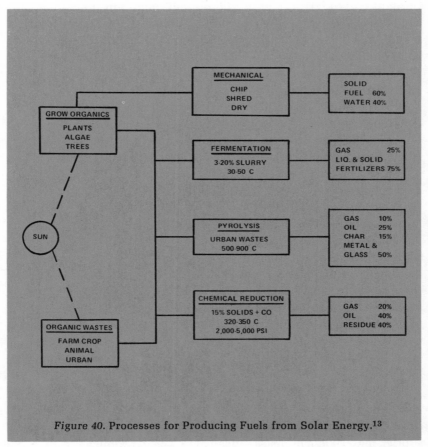

Figure 40. Processes for Producing Fuels from Solar Energy.[13]

fuel would be $0.06/MBTU for a $250/acre land cost, 1400 BTU/ft^2/
day insolation, 3% capture efficiency, 8% interest rate, 0.6% tax
rate, and $200/acre harvesting cost, and the total cost of the electric
power from this "energy plantation" is computed to be 5 mills/
kWh, based on a $200/kW capital cost and 28-year life for the
power plant.[100] A "worst case" fuel cost was determined to be
$0.40/MBTU if the capture efficiency is reduced to 1% and the
harvesting cost increased to $700/acre, which results in a power
cost of 8.5 mills/kWh. The annual operating cost is taken to be $2
million/year, and insurance and tax costs 0.12% and 2.35% of the
capital cost of the power plant. This study concluded that this type

of plant would cost no more to build and maintain than a conventional fossil fuel steam-electric plant and that the energy plantation is a renewable resource and an economical means of harnessing solar energy. It is not at all obvious at the present time what type of plant (trees, grasses, etc.) will result in the lowest power costs. The NSF/NASA Solar Energy Panel concluded that using trees the fuel cost at the power plant might range from $1.50 to $2/MBTU.[13]

Some power can also be produced by the combustion of organic wastes, which also reduces problems of disposal of these wastes. It has been estimated that the total animal and solid urban wastes which can be collected at reasonable cost could provide about 6% of the heat energy requirements for electric generating plants. The most promising use of solid animal wastes is in connection with large feedlot operations where large quantities are accumulated at one location and disposal presents a continuing problem.

Anaerobic fermentation of organic materials results in the production of methane and carbon dioxide. This process can be used (Figure 41) to convert from 60% to 80% of the heating value of organic materials into methane, which can serve a wide variety of uses including powering automobiles. Methane can also be used in existing natural gas pipelines. Algae grown in sewage ponds can be used for the production of methane; costs of producing methane by this method are estimated between $1.50 and $2/MBTU.[13]

Pyrolysis has also been used for many years to convert organic materials to gaseous, liquid and solid fuels. Any organic materials can be used, and in addition plastics, rubber products, and other similar materials can also be used. The gases produced are a mixture of hydrogen, methane, carbon monoxide, carbon dioxide, and hydrocarbons. About two barrels of oil can be produced per ton of dry organic material. A plant handling 1000 tons of waste per day (Figure 42) could dispose of the solid wastes produced by a city of 600,000 people.

At temperatures around 600°F and pressures between 2000 psi and 4000 psi organic materials can be partially converted into oil. In laboratory tests oil yields up to 40% of the weight containing about two-thirds of the heating value of the initial dry organic matter have been obtained.

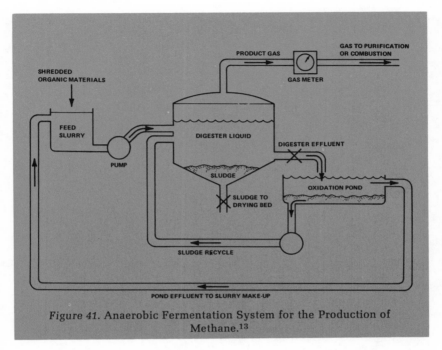

Figure 41. Anaerobic Fermentation System for the Production of Methane.[13]

Hydrogen may be thermochemically produced directly from water using solar heat. For example, one regenerative chemical cycle operates with bromides of calcium and mercury in a four step process with a maximum temperature of 1350°F.[102] The four reactions are

1) $CaBr + 2H_2O$ $\rightarrow$ $Ca(OH)_2 + 2HBr$ 1350°F
2) $Hg + 2HBr$ $\rightarrow$ $Hg\,Br_2 + H_2$ 480°F
3) $HgBr_2 + Ca(OH)_2$ $\rightarrow$ $CaBr_2 + HgO + H_2O$ 400°F
4) HgO $\rightarrow$ $Hg + (0.5)O_2$ 900°F

The net result of these four reactions is:

$$H_2O \rightarrow H_2 + (0.5)O_2$$

Water is thus separated into hydrogen and oxygen at temperatures easily obtainable by linear concentrators. The hydrogen and oxygen are released at separate points in the cycle, and the chemicals used are regenerated permitting virtual 100% recovery of the chemicals

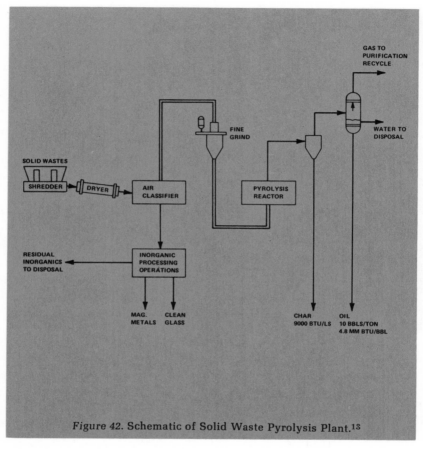

Figure 42. Schematic of Solid Waste Pyrolysis Plant.[13]

without sideloops. One drawback of this particular regenerative process is the large amount of materials circulated per unit product. This cycle is an example of a large number of regenerative thermochemical cycles that have been proposed for the production of hydrogen with temperatures obtainable on a large scale with solar concentrators.[103]

Figure 43 illustrates the relative 1972 cost of solar-produced clean renewable fuels and fossil fuels. The costs of fossil fuels have risen considerably since 1970, so solar synthetic fuels are going to continue to become increasingly competitive.

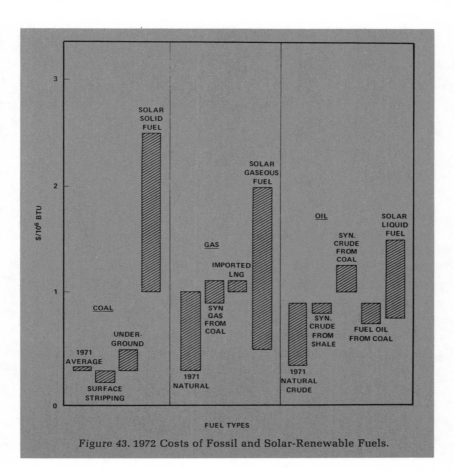

Figure 43. 1972 Costs of Fossil and Solar-Renewable Fuels.

Chapter 12

OCEAN-THERMAL POWER

The French physicist Jacques D'Arsonval suggested in 1881 that a heat engine operating between the warm upper layer and the cold deep water of the tropical oceans could produce large amounts of power.[104] Although the engine must be inherently inefficient, the amount of heat available is enormous, and since this heat comes from the sun, ocean-thermal power is appropriately classified as a form of solar power. D'Arsonval suggested a number of possible high vapor pressure working fluids, including ammonia.

In 1929 Georges Claude, a friend of D'Arsonval, demonstrated a 22-kilowatt ocean-thermal power plant in Mantanzas Bay, Cuba (Figure 44), but due to its low efficiency (<1%) the plant was not economically competitive with other power plants at that time.[105] Claude used surface sea water admitted to a low-pressure evaporator to provide low-pressure steam to drive the turbine. This low-pressure steam was then recondensed by direct contact with cold seawater in a spray condenser. The Claude cycle avoided large heat exchangers required by closed-cycle plants to vaporize and recondense a high vapor pressure working fluid, but did require a large turbine of inherently low efficiency. The relatively high vacuum required maintaining large leak-tight connections and the removal of dissolved gases from the water. The plant itself was located on land and 2-km long tubes brought cold water from the depths, with resulting heating of the water as it flowed through the tubes. In spite of the economic failure of the project, Claude's plant was the first to demonstrate power generation from ocean temperature gradients.

Two large experimental power plants of 3.5 MWe each using the

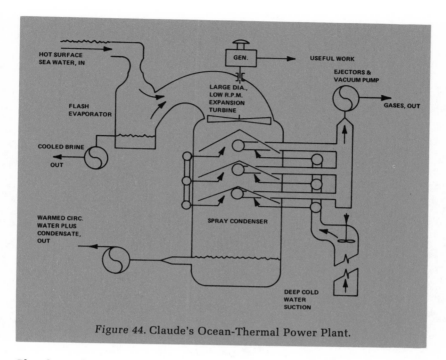

Figure 44. Claude's Ocean-Thermal Power Plant.

Claude cycle were built by the French at Abidjan off the Ivory Coast in 1956 to utilize a thermal difference of 36°F. An 8-ft-diameter pipeline was built extending to a depth of 3 miles about 3 miles from shore, but difficulties in maintaining this pipeline prevented the plant from operating at full capacity. About 25% of the power generated was required for the pumps and other plant accessories. The plants were finally abandoned.

Two approaches to improving the Claude cycle are the use of controlled flash evaporation[106] and the indirect vapor cycle.[107] The controlled flash evaporation system (Figure 45) eliminates major problems of deaeration and seawater corrosion associated with the Claude cycle and produces fresh water in addition to electric power. The flash evaporator consists of a large number of parallel vertical chutes with films of warm seawater flowing down. As the pressure drops, water evaporates and the vapor flows downward. This low-pressure steam then flows through the large low-pressure turbine and into the condenser where it is cooled and condensed by cold

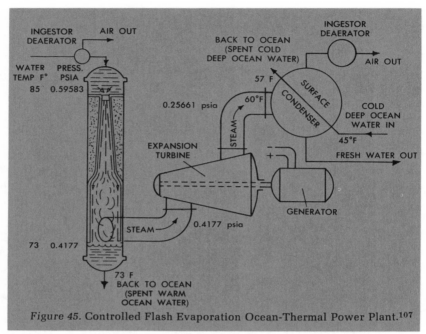

Figure 45. Controlled Flash Evaporation Ocean-Thermal Power Plant.[107]

seawater from the ocean depths. If fresh water is not desired, steam from the turbine can be condensed by direct contact with cold sea-water (as in the Claude cycle) with a slight increase in power output. Deaeration in this cycle is accomplished at a low cost with practically no power requirement. About 11.5 gallons of pure water can be produced per 1000 gallons of warm water circulated. This system still suffers from the large, inherently inefficient low-pressure steam turbine.

The indirect vapor cycle requires the addition of a boiler, but permits the use of higher-pressure working fluids with a much smaller and more efficient turbine (Figure 46). Since the efficiency of ocean thermal plants can be only about one-tenth that of modern steam plants, the amount of heat transferred in the boiler and condenser per unit power output must be about ten times as large. It does not follow, however, that the costs of these components will be ten times as great. Since ocean-thermal plants will operate at relatively low pressures and ambient temperatures, the tube walls can be thinner and cheaper materials can be used, so the cost per unit of

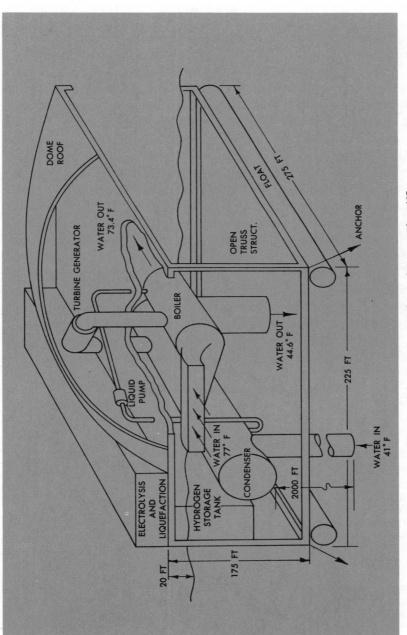

Figure 46. Floating Ocean-Thermal Power Plant.[107]

heat transfer should be much less for ocean thermal boilers and condensers than for those used in high-temperature steam plants.

A floating power plant has been proposed that could use propane as the working fluid in an indirect vapor cycle. Seawater from the warm surface layer is passed through the boiler to vaporize propane at about 150 psi. The propane exhausted from the turbine is condensed at about 110 psi by cold seawater. In 1965 the capital cost of this plant was estimated at $168/kW, which was comparable to the capital cost of a fossil-fueled plant at that time.[108] To equalize pressure differences in the boiler and condenser, the plate heat exchanger acting as the boiler could be lowered to a depth of 290 ft and the plate condensers lowered to 150 ft, with the turbines and other components at intermediate depths. A modular design has been suggested with the boiler, condenser and engine modules all of the same standard size, such as 8 ft × 8 ft × 40 ft (Figure 47), which should reduce manufacturing, transportation, and assembly costs. The plant would be neutrally buoyant at the depth which minimizes the pressure differences in the boiler and condenser.[109]

Characteristics of potential working fluids are given in Table 14.[110] The ideal cycle efficiency is based on a maximum cycle

Table 14. Comparison of Working Fluids[110]

Fluid	Ideal Cycle Efficiency (%)	Cycle Efficiency (5% ΔP/P)	High Pressure (psia)	Low Pressure (psia)	Pump Work (kW)	Ideal Mass Flow (lb/min)
Ammonia	3.72	2.71	118	81	1079	317,600
Butane	3.82	2.81	29	20	859	976,000
Carbon Dioxide	2.89	1.67	799	609	36,033	2,873,000
Ethane	3.90	2.04	53	411	25,300	1,495,000
R-12	3.68	2.57	78	56	2,450	2,630,000
R-22	3.68	2.54	126	91	3,200	1,978,000
R-113	3.65	2.91	5	3	170	2,436,000
R-500	3.67	2.55	92	66	2,750	2,205,000
R-502	3.61	2.41	140	103	4,552	2,756,000
Propane	3.67	2.46	115	85	3,706	1,084,000
Sulphur Dioxide	3.72	2.82	45	30	634	1,041,000
Water	3.78	3.26	0.3	0.15	1.4	155,500

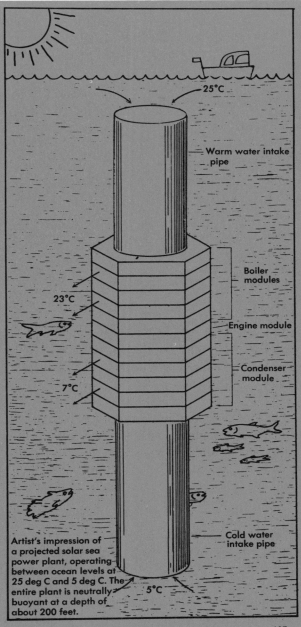

25°C

Warm water intake pipe

Boiler modules

23°C

Engine module

Condenser module

7°C

Cold water intake pipe

Artist's impression of a projected solar sea power plant, operating between ocean levels at 25 deg C and 5 deg C. The entire plant is neutrally buoyant at a depth of about 200 feet.

5°C

Figure 47. Modular Ocean-Thermal Power Plant.[107]

temperature of 65°F and a minimum cycle temperature of 45°F. Ammonia is the best working fluid from the heat transfer standpoint (Table 15).

Table 15. Heat Transfer Coefficients of Working Fluids[110]

Fluid	Relative (Condensing)	Relative (Boiling)
Ammonia	1	1
Butane	0.15	0.32
Carbon Dioxide	0.18	0.18
Ethane	0.11	0.21
R-12	0.11	0.11
R-22	0.16	0.14
R-113	0.09	0.13
R-500	0.12	0.13
R-502	0.10	0.11
Propane	0.13	0.27
Sulphur Dioxide	0.38	0.33
Water	0.92	2.27

For ammonia a single-stage turbine with a 7-ft wheel diameter could generate 25 MW at 1800 rpm; for propane a 12-ft wheel diameter single-stage turbine could produce 30 MW at 600 rpm. Propane and ammonia appear at present to be the most attractive working fluids.

The amount of energy available for ocean-thermal power generation is enormous, and is replenished each year as the sun heats the surface layers of oceans and melts snow in the arctic regions causing cold currents to flow deep beneath the surface toward the equator. It has been estimated that the tropical oceans in the year 2000 could supply the whole world with energy at a per capita rate of consumption equal to the U.S. per capita rate in 1970 and suffer only a one-degree C drop in temperature. Also, if nutrient-rich cold water is brought from the ocean depths and released near the surface, this could result in a substantial increase in fish populations, as occurs naturally off the coast of Peru. Another advantage could be a slight lowering of tropical temperatures. At a depth of 1300 ft 30 miles from Miami the temperature is 43°F, as compared with a surface temperature from 75°F–84°F, so this could be a good location for an ocean-thermal power plant.

A "sea plant" has been proposed with a floating propane cycle ocean-thermal electric power plant, a separate ocean-thermal flash evaporation plant for producing pure water, and various chemical industries based on extracting oxygen and raw materials from the ocean.[111] Noting that the Gulf Stream alone could supply 200 times the total power requirements of the United States, the cost of a 100-MWe plant is estimated at $20 million ($200/kWe) and the cost of fresh water at $0.04/1000 gallons. This cheap power and cheap water makes possible a variety of energy intensive chemical process plants. Oxygen gas, extracted from seawater, could be liquefied using propane turbines to drive the refrigeration compressors, and cold water from the ocean can be used as a convenient heat sink at lower than usual ambient temperatures. Chemical plants using raw materials extracted from seawater would benefit from the cheap power. Bromine and magnesium are already being produced commercially from seawater.[112] In addition, one of the best ways to transmit power to shore may be to electrolyze water to produce hydrogen and oxygen, and then liquefy these gases, which can then be shipped or piped to shore.

Although fossil power plants tend to cost less, the cost of delivered power can be higher because of fuel costs, which are negligible for solar, sea-thermal, geothermal, and wind plants. These cost ranges are subject to considerable change as technology advances and economic conditions change.

Another possible sea-thermal conversion system, which does not use any working fluid, is the Nitinol engine developed by researchers at the Lawrence Berkeley Laboratories. The Nitinol wires bend and straighten as they are immersed in cool and warm water, respectively. The force they exert during this process is converted into rotational motion, and small amounts of electric power have been generated by a laboratory model. Since this engine operates between temperature differences of only a few degrees it may eventually be useful for sea-thermal power systems. The small laboratory model is reported to operate at 70 rpm with a temperature difference of 41°F to produce 0.23 watts with an absolute thermal efficiency of 3.4%.[113]

Chapter 13

GEOSYNCHRONOUS POWER PLANTS

The concept of placing a large solar array in geosynchronous orbit and transmitting this power to earth was proposed in 1968,[113, 114] and since has received increasing attention as a potential major energy resource for the next century. The basic motivation for placing a solar array in space is the increased availability of solar energy in space, as illustrated by Table 16. Up to 15 times as much solar

Table 16. Average Availabilities of Solar Energy[115]

Availability Factor	Average On Earth	In Synchronous Orbit	Average Ratio
Solar Radiation Energy Density	0.11 watts/cm²	0.14 watts/cm²	4/5
Percentage of Clear Skies	50%	100%	1/2
Cosine of Angle of Incidence	0.5	1.0	1/2
Useful Duration of Solar Irradiation	8 hr	24 hr	1/3
Product			1/15

energy is received by a solar array in space as the same array would receive on the ground, and this energy is received continuously, nearly 24 hours a day. Now that NASA is developing the space shuttle to permit the routine exploitation of the space surrounding the earth, the economics of geosynchronous power plants are becoming more attractive.

The basic concept is illustrated by Figure 48. Concentrators would reflect sunlight onto an advanced, lightweight solar array.

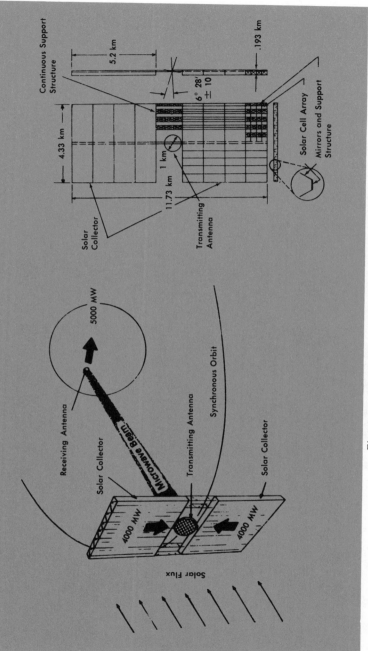

Figure 48. Geosynchronous Solar Power Plant.[116]

The two symmetrically arranged collectors convert solar energy directly to electricity which powers microwave generators within the transmitting antenna located between the two large collecting panels. The 1-km-diameter antenna transmits the power to a 7.4-km-diameter receiving antenna on the ground (Figure 49) with an over-

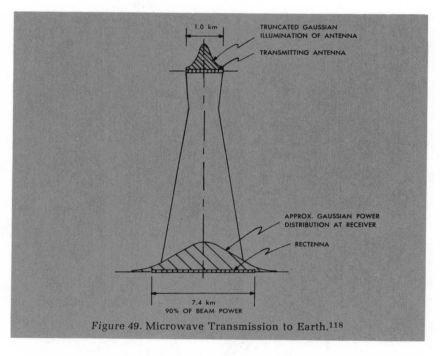

Figure 49. Microwave Transmission to Earth.[118]

all efficiency of about 68%. The microwave transmission system is expected to cost about $130/kWe.[117] To achieve the necessary coherent transmission, the many separate elements of the transmitting antenna are phase-locked onto a pilot signal originating from the center of the receiving grid, so that it is impossible to direct the beam away from the receiving antenna. Since the receiving grid does not block sunshine, the land beneath can be used for growing farm crops. Microwave intensities reaching the earth are safe.

The solar cells in the array are projected to have an 18% efficiency, 2-mil thickness, and cost $0.28 per cm², which should lead to a 430-watt/lb array costing $0.68 per cm² and having a 30-year life. The array is expected to suffer a 1% loss of solar cells from micrometeoroid impacts over a 30-year period. Cost estimates for a small several hundred-megawatt prototype plant, based on current shuttle cost estimates and near-term solar cell technology, are given as $310/kWe for the solar arrays, $230/kWe for the microwave transmission system, and from $800/kWe to $1380/kWe for transportation to geosynchronous orbit and assembly, for a total system cost of from $1340/kWe to $1920/kWe. Capital cost for a fully operational 5000 MWe plant is expected to be about $800/kWe. The power satellite will produce more energy in its first year of operation than was required to manufacture it and place it in orbit.[117]

Large geosynchronous solar-thermal plants (Figure 50) operating with a "current technology" helium/xenon brayton cycle have also been considered. The capital cost of a 1980 technology plant is estimated at $2540/kWe[119] Since about 80% of this cost is space

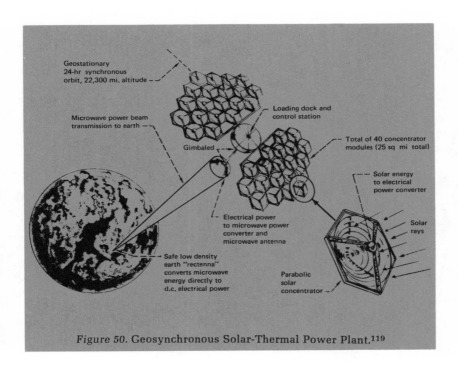

Figure 50. Geosynchronous Solar-Thermal Power Plant.[119]

transportation, this cost should be reduced if a fully reusable space shuttle becomes operational and lighter-weight reflecting surfaces become available. Based on the same projected 1980 technology, advanced solar cell systems are estimated to cost $2950/kWe, slightly more than the solar-thermal system. Another estimate for the capital cost of solar cell geosynchronous plants that was given lies in the range of $1400/kWe to $2600/kWe.[120] An earlier estimate of $2100/kWe was given for a prototype solar cell synchronous power plant based on a study by the A. D. Little/Grumman/Raytheon/Textronics team.[121] This group has been conducting a study of the solar cell synchronous power satellite for several years. Another major study of solar-thermal and nuclear power satellites has been conducted by Boeing, and it showed that an ambitious power satellite program could supply a major portion of our nation's electric power needs between 1990 and 2020 and could pay for itself.

Chapter 14

POWER FROM THE WIND

A small fraction of the solar energy falling on the earth each day is converted into surface winds that are quite strong and in some areas provide a useful source of energy for performing mechanical work and generating electric power. Although windmills have been used for more than a dozen centries for grinding grain and pumping water, interest in large-scale electric power generation has developed over the past 50 years. In 1920 J.B.S. Haldane wrote: "Ultimately we shall have to tap these intermittent but inexhaustible sources of power, the wind and sunlight. The problem is simply one of storing their energy in a form as convenient as coal or petrol. . . . During windy weather the surplus power will be used for the electrolytic decomposition of water into oxygen and hydrogen. These gases will be liquefied and stored in vats, vacuum jacket reservoirs, probably sunk in the ground."[122]

In 1939 work was begun on a 1.25-MWe wind power plant on Grandpa's Knob near Rutland, Vermont. Electricity was generated and delivered to the utility transmission grid in October, 1941, the first synchronous generation of power from the wind. The rotors and electric generator were mounted on a 110-ft tower and turned in any direction to face the wind. The two stainless steel blades weighed 7.5 tons each and swept a circle 175 ft in diameter with a rated speed of 28.7 rpm. Full power operation was achieved for wind velocities in excess of 30 miles per hour, which occurred about 70% of the time. Icing of the blades was not a problem since the ice would break up during rotation. The total weight of the wind power generator was 250 tons, and the cost was slightly over one million dollars.[123]

Partly because the project was rushed to completion in the days preceding World War II, it was plagued with component failures. Replacements were especially difficult to obtain during the war. Finally, on March 26, 1945, the blade broke during a storm. Because of the limited financial resources of the company operating the plant the generator was not repaired, but dismantled and removed from the site.[124]

Based on the experience at Grandpa's Knob, the Federal Power Commission conducted a study of wind-electric power generation for use with interconnected utility networks, and concluded that a power plant capacity between 5 and 10 megawatts could make wind power economical. A 7.5-MWe unit was designed using 2-bladed propellers, similar to the propellers used on small airplanes. A separate design for a 6.5-MWe plant used 3-bladed propellers. A wind-driven dc generator provided power to a converter which produced synchronous ac power. The projected costs of these plants, in 1945 dollars, were $68 per kilowatt of capacity for the 7.5-MWe unit and $75 per kilowatt for the 6.5-MWe unit.[125]

In addition to the Grandpa's Knob experiment, a variety of similar projects have been undertaken around the world. A 100-kWe direct current wind turbine with a 30-m propeller diameter was operated in the Soviet Union in 1931. In 1942 a 3-bladed propeller 50-kWe ac plant was operated in Germany, and the next year a 20-kWe generator using two 6-bladed 9-m-diameter propellers was tested in Berlin. In Denmark, a 200-kWe generator with a single 3-bladed propeller of 24-m diameter has been operated, and a 100-kWe 15-m 3-bladed turbine has been tested in England. Between 1961 and 1966 a 35-m diameter, 100-kW double-bladed wind turbine operated in West Germany, following tests with a 10-kWe model. The power output of the 100-kWe unit increased linearly from 10 kWe at a wind velocity of 4 m/sec to 90 kWe at 9 m/sec. The power output was usually held to 90 kWe for wind velocities greater than 9 m/sec. The most spectacular European wind generator has been the 31-m-diameter, 800-kWe generator built in France in 1958. This generator used a single 3-bladed propeller.[126]

At present a company in Switzerland is manufacturing 5-kilowatt wind generators with a 5-m-diameter propeller at a cost of $1900, plus $200 freight for delivery to the United States, with delivery in about 6 weeks. A 400-watt generator is also on sale in Germany, and a 1-kWe generator is made in Italy.[127]

One system for a house in Maine uses a 2 kWe wind generator manufactured in Australia with 19 storage batteries and a small dc-ac inverter to provide all the electric power needs for the home, including power for lights, household appliances, power tools, and the television.[128] The storage batteries provide enough reserve power for four days without wind; a gasoline generator is used as an emergency backup system in case of prolonged calm periods. The only maintenance required for this system is a change of one quart of oil in the gearbox once every five years.

The wind generator uses a 12-ft diameter propeller and produces up to 2 kWe of 115-volt dc power. The voltage regulator panel is incorporated as part of the wind generator. The transistorized voltage control lowers the charging rate when the output voltage exceeds the voltage of the batteries in the fully charged state. An anemometer on the wind generator measures the wind speed in miles per hour. The 19 130-amp-hour batteries are connected in series to provide 15 kWH of storage at 115 volts. The lights and many of the appliances including the vacuum cleaner, electric drill, skill saw, sewing machine, and water pump are operated on dc. Only the television and stereo require ac, which is provided by a small inverter.

This system costs about $2800, including $1600 for the wind generator, $700 for the batteries, $100 for the small inverter, and $200 for wiring and other small components. The cost of the electric power for this home is about 15 cents per kWH—about half the cost of power from a gasoline or diesel plant, but much higher than the power company rate. This house was located 5 miles from the nearest paved road and the local power company would have charged $3000 to run a line to the house, so the wind power system was the most economical source of electric power for this home.

For those requiring more power, a company in Switzerland sells a 6-kWe wind generator complete with automatic controls, for

$3000.[128] This unit has a 3-bladed propeller of 16.5-ft diameter and can operate completely automatically and unattended in winds as high as 150 mph. Full output is delivered in a 25-mph wind. Solid state inverters are also available; for example a 2-kWe dc-ac inverter sells for about $1600.

For commercial electric power generation, wind generators of power outputs in excess of 100 kWe will be needed. No storage is required for the utility network as long as the average hydroelectric generation is several times greater than the power provided by wind generators, because the output of the hydroelectric plants can be varied to compensate for changes in wind power output. Also, the use of large numbers of wind generators at separate sites will tend to reduce storage requirements and smooth out short-term fluctuations in total wind power output.

The largest wind generator built in recent times was the 800-kWe unit operated in France from 1958-1960. The flexible 3-bladed propeller was 105 ft in diameter and produced the rated power in a 49-mph wind with a rotational speed of 47 rpm. The maximum power delivered was 1.2 MWe. After 18 months in operation the flexible propeller was replaced by a rigid one in an effort to improve the performance of the generator. One of the blades of the new propeller soon broke and the hub of the machine was torn up by the unbalanced forces acting on the propeller. The generator was never operated again.[129]

A 100-kWe generator will soon be built by NASA with NSF support at the Lewis Research Center Plum Brook test facility at Sandusky, Ohio. This will be the largest wind generator built in the United States since the Grandpa's Knob experiment more than thirty years ago. The project will study the performance, operating characteristics, and economics of wind systems for the generation of commercial electric power. The metal rotor blades will be located on the downwind side of the tower. The generator is expected to produce 100 kWe output at 40 rpm at a wind speed of 18 mph, and should generate 180,000 kWH/year in the form of 460-volt, 3-phase, 60-cycle ac power.[130]

If both wind generators and solar-thermal or photovoltaic generators become commercially viable, using both in a utility power network will be a definite advantage since the availability of sunlight seldom coincides with the availability of wind. Winds are often most vigorous in cloudy weather when photovoltaic array outputs would be minimal and solar-thermal power systems would not work at all. On the other hand, on clear, calm days plants using sunlight would function best and wind plants not at all. If a large number of generators using sunlight and wind are dispersed over a large geographical area but connected to the *same power grid,* storage requirements would be minimal.

For small applications, solar heat and wind power systems complement each other well, because sunlight is most useful at present for providing heat and wind generators for producing power. Wind can now be used to meet electrical power needs while solar energy is used for heating and cooling. The Grumman Aerospace Corporation has recently begun to market its Windstream-25 wind power system at a cost of about $25,000. This system includes a 25-ft diameter wind generator on a 40-ft tower with a capacity of 18 kWe in a 26-mph wind, automatic controls, battery storage, and ac conversion.

Chapter 15

SOLAR ENERGY TOMORROW

Over millions of years ago plants covered the earth, converting the energy of sunlight into living tissue, some of which was buried in the depths of the earth to produce deposits of coal, oil and natural gas. During the past few decades man has found many valuable uses for these complex chemical substances, manufacturing from them plastics, textiles, fertilizer and the various end products of the petrochemical industry. Each decade sees increasing uses for these products. Coal, oil, and gas are nonrenewable natural resources which will certainly be of great value to future generations, as they are to ours.

However, man has found another use for these valuable chemicals from the earth—a use other than the creation of the products that add so much to our standard of living. That is to burn them, to burn them in huge and ever-increasing quantities to power the machines of society and provide heat. They are being burned at such an incredible rate that in a few short decades the world reserves of natural gas may be depleted, decades later the oil will be gone, and in a century or two the world will also be without coal. Undoubtedly successive generations after that time will decry the excesses of the present generation in selfishly destroying these valuable resources without regard for the welfare of their descendants.

It should now be clear to the reader that the rapid depletion of nonrenewable fossil resources need not continue, since it is now or soon will be technically and economically feasible to supply all of man's energy needs from the most abundant energy source of all, the sun. Sunlight is not only inexhaustible, it is the only energy source which is completely nonpolluting. The land area required to provide all our energy is a small fraction of the land area required to produce our food, and the land *best* suited for collecting solar energy (rooftops and deserts) is the land *least* suited for other purposes. It is time for the United States, which led the world in the development of atomic energy and putting men on the moon, to mount an equally massive effort to usher in the Solar Age. A massive effort by our country can offer the world the technology for the economical utilization of solar energy in all its varied forms—photovoltaic, direct solar-thermal, renewable fuels, ocean-thermal, and wind. Then we can conserve our valuable nonrenewable fossil resources for future generations to enjoy, and we can all live in a world of abundant energy without pollution.

Appendix A. Conversion of Units

To convert from	to	multiply by
acre	meter^2	4046.856
acre-feet	gallon	2.36×10^5
acre-feet	meter^3	1233.5
angstrom	meter	1×10^{-10}
angstrom (Å)	microns	1×10^{-4}
atmosphere	newton/meter^2	101,325
atmosphere	lb/in^2	14.70
bar	newton/meter^2	10^5
bar	atm	0.9869
bar	lb/in^2	14.5038
board foot (1' x 1' x 1")	meter^3	0.0023597
British thermal unit	joule	1054.35
BTU	cal	251.996
BTU	kWh$_t$	2.928×10^{-4}
BTU/h	cal/sec	0.0700
BTU/h	Watts	0.2931
BTU/lb	J/kg	2.326×10^{-3}
BTU/lb	cal/g	0.55555
BTU/ft^{2}h	W/m^2	3.1524
BTU/ft^{2}F$^\circ$h	cal/m^{2}C$^\circ$sec	1.3573
BTU/ft^{2}F$^\circ$h	W/m^{2}C$^\circ$	5.6783
calorie	joule	4.184
calorie	BTU	3.9683×10^{-3}
cal/g	BTU/lb	1.80000
cal/cm^{2}sec	BTU/ft^{2}h	1.3272×10^4
cal/cm^2	L (langley)	1.0000
Celsius (temperature)	kelvin	$t_k = t_c + 273.15$
centimeter	inch	0.39370
centimeter	feet	0.03281
cm^2	in^2	0.15500
cm^2	ft^2	1.0764×10^{-3}
cm^3	in^3	6.1023×10^{-2}
cm^3	ft^3	3.5315×10^{-5}
centimeter of mercury	newton/meter^2	1333.22
centimeter of water	newton/meter^2	98.0638
cubic feet per minute	liters/sec	0.4720
day (mean solar)	second (mean solar)	86400
day (sidereal)	second (mean solar)	86164
dyne	newton	10^5
erg	BTU	9.4845×10^{-11}
erg	joule	10^{-7}
Fahrenheit	kelvin	$t_k = (5/9) (t_f + 459.67)$
Fahrenheit	Celsius	$t_c = (5/9) (t_f - 32)$

To convert from	to	multiply by
feet	centimeter	30.480
feet	meter	0.30480
feet of water	lb/in^2	0.4335
feet^2	cm^2	929.03
feet^3	cm^3	2.8317×10^4
feet^3	m^3	0.028317
feet/second	cm/sec	30.48
feet/second	miles/hour	0.68182
foot	meter	0.3048
feet of water	newton/meter^2	2989
footcandle	lumen/meter^2	10.764
gallon (U.S.)	meter^3	0.0037854
gallon (U.S.)	cm^3	3785.41
gallon (U.S.)	pound (water)	8.34517
gallon (U.S.)	kg (water)	3.7852
gallon (U.S.)	ft^3	0.13368
gallon (U.S.)	acre-feet	3.0689×10^{-6}
gallon (U.S.)	liters	3.7852
g	pound	2.2046×10^{-3}
g/cm^2	pound/inch	1.4223×10^{-2}
g/cm^2	atm	0.6784×10^{-4}
hectare	meter^2	10^4
horsepower	watt	746
horsepower	BTU/min	42.4356
horsepower	kW	0.74570
hour (mean solar)	second (mean solar)	3600
hour (sidereal)	second (mean solar)	3590.17
inch	meter	0.0254
inch	cm	2.54000
inches of water	lb/in^2	0.036126
inches of water	atm	2.458×10^{-3}
inch^2	cm^2	6.4516
inch^2	feet^2	6.9444×10^{-3}
J/sec	BTU/minute	0.056907
j/sec	hp	1.3410×10^{-3}
kJ	kWh	2.7778×10^{-4}
kJ/m^2	BTU/feet^2	0.0088111
kJ/m^2	cal/cm^2	0.023901
kJ/m^2	kcal/m^2	0.23901
kg	pound	2.20462
km	miles	0.62137
km	feet	3280.84
kcal	J	4184
km^2	mi^2	0.38610
kW	BTU/h	3414.43
kW	hp	1.3410
kWh	BTU	3410.08

To convert from	to	multiply by
kWh	cal	8.59326×10^5
kWh	kcal	8.59326×10^2
kW/m^2	BTU/ft^{2}h	317.21
kilocalorie	joule	4184
kilogram force (kgf)	newton	9.80665
langley	BTU/ft^2	3.6866
langley	joule/meter^2	41840
lbf (pound force)	newton	4.44822
liter	in^3	61.0254
liter	gallon	0.26418
liter	ft^3	3.53157×10^{-2}
lumen	W	1.4706
meters	miles	6.2137×10^{-4}
m	feet	3.2808
m	inch	29.3701
m^2	feet^2	10.7639
m^2	ha	10^{-4}
m^2	acres	2.47105×10^{-4}
MBTU	kJ	1.0548×10^6
mi	yards	1760
mi^2	km^2	2.58999
micron	meter	10^{-6}
mile	meter	1609.34
minute (mean solar)	second (mean solar)	60
month (mean calendar)	second (mean solar)	2628000
pound	kg	0.453592
Rankine	kelvin	$t_K = (5/9)t_R$
therm	BTU	10^5
therm	MBTU	0.10000
therm	kWh	29.28
ton (metric)	kilogram	1000
ton (short, 2000 lbs)	kilogram	907.1847

Appendix B. Circulating Pumps for Hydronic Solar Systems

This appendix provides information on the various types of circulating pumps available for solar systems that use water or an antifreeze solution as the heat transfer fluid. This information includes manufacturer, type of pump (model number), motor horsepower (1/25 to 3), motor RPM, pump curves (head in feet versus gallons per minute), pump material (iron, bronze, stainless steel), maximum operating temperature, pump connections (flanged, national pipe thread), and pump cost. Applicable pumps are those that can pump water or an antifreeze solution and operate at a high temperature. Bronze or stainless steel pumps are rated for circulating potable water, but cast iron pumps are not.

With these requirements and the manufacturer's information in mind, Table B-1 was constructed. It lists the pump manufacturers in alphabetical order, with all pertinent pump information about each specific model included.

Also, for each horsepower rating, a plot of the pump characteristics was made for the various manufacturers and various models on the same graph. For example, all 1/2-horsepower pumps from different companies and different models were plotted on the same graph to permit immediate comparison of the various pump options at 1/2 horsepower. This was done for the horsepower range from 1/25 to 3 horsepower. Each curve on each figure was given a number and corresponds to a specific pump found on Table B-1. By using the figures in conjunction with Table B-1 the appropriate circulating pump for a particular hydronic solar system may be chosen.

Table B-1

Manufacturer and Other Information	Curve No.	Type	HP	RPM	Iron	Bronze	Stainless Steel
						Cost ($)	
Armstrong: Max. Temp.—225°F, Flanged Conn. S—Standard, H—High Duty							
	1	S-25	$\frac{1}{12}$	1750	73	115	
	2	S-34	$\frac{1}{6}$	"	103	167	
	3	S-35	$\frac{1}{6}$	"	122	205	
	4	H-32	$\frac{1}{6}$	"	105	170	
	5	H-41	$\frac{1}{6}$	"	138	199	
	6	S-45	$\frac{1}{4}$	"	174	327	
	7	S-46	$\frac{1}{3}$	"	239	402	
	8	S-55	$\frac{1}{2}$	"	250	421	
	9	S-57	$\frac{3}{4}$	"	306	458	
	10	S-69	1	"	440	689	
Aurora: Max. Temp.—225°F, NPT Conn.							
	11	321 1-1 & 1/4-4	$\frac{1}{3}$	3500	190		
	12	321 1-1 & 1/4-4	$\frac{1}{2}$	"	213		
	13	321 1 & 1/2-2-4	$\frac{3}{4}$	"	237		
	14	321 1 & 1/2-2-4	1	"	256		
	15	321 1 & 1/2-2-4	1 & $\frac{1}{2}$	"	321		
	16	321 2-2 & 1/2-5	3	"	536		
Bell & Gossett: Max. Temp.—225°F, Flanged or NPT Conn.							
	17	Series 100	$\frac{1}{12}$	1750	75	122	
	18	Series PR	$\frac{1}{6}$	"	145	203	
	19	Series HV	$\frac{1}{6}$	"	114	172	
	20	Series 1 & 1/2	$\frac{1}{6}$	"	112	166	
	21	2 & 1/2	$\frac{1}{4}$	"	183	333	
	22	60-11	$\frac{1}{4}$	"	176		
	23	HD3	$\frac{1}{3}$	"	249	431	
	24	PD35-S	$\frac{1}{2}$	"	263	437	
	25	60-13	$\frac{1}{2}$	"	217		
	26	PD37-S	$\frac{3}{4}$	"	325	475	
	27	60-14	$\frac{3}{4}$	"	240		

Table B-1. Continued

Manufacturer and Other Information	Curve No.	Type	HP	RPM	Cost ($)		
					Iron	Bronze	Stainless Steel
Bell & Gossett: (Cont.)	28	PD38-S	1	1750	425	634	
	29	60-17	1	"	405		
	30	101-T	1	"	521		
	31	PD40-S	1 & ½	"	520	730	
	32	102T	1 & ½	"	532		
	33	103T	2	"	600		
	34	104T	3	"	615		
Buffalo Forge: Max. Temp.—225°F, Flanged Conn.	35	AA6	1	3500	566		
	36	AA6	1 & ½	"	566		
	37	AA6	2	"	577		
	38	AA6	3	"	590		
Dean Line: Max. Temp.—220°F, Flanged Conn.	39	3/4 DL	½	1750	331		503
	40	1 & 1/2 DL	½	1750	361		610
	41	3/4 DL	1	3500	339		520
	42	1 & 1/2 DL	1	3500	370		624
	43	3/4 DL	1 & ½	"	350		546
	44	1 & 1/2 DL	1 & ½	"	391		636
	45	3/4 DL	2	"	374		570
	46	1 & 1/2 DL	2	"	419		663
Enpo Cornell: Max. Temp.—212°F, Flanged or NPT Conn.	47	CC 1 & 1/4 J-A	¼	1800	239		
	48	CC 1 J	½	3600	237		
	49	CC 1 & 1/4 J	¾	3600	235		
	50	CC 1 & 1/4 J	1	3600	243		
	51	CC 1 & 1/4 J	1 & ½	3600	271		

Fairbanks Morse:
Max. Temp.—200°F, NPT Conn.
P—Nylon Impeller

52	1 C2J	⅓	3450	186
53	1 C2JP	⅓	"	154
54	1 C2J	½	"	193
55	1 C2JP	½	"	169
56	1 C2J	¾	"	225
57	1 C2JP	¾	"	179
58	1 & 1/2 C1J	1	"	252
59	1 & 1/4 C2J	1	"	230
60	1 & 1/4 C2JP	1	"	204
61	1 & 1/2 C1J	1 & ½	"	288
62	2 C1J	1 & ½	"	316
63	1 & 1/4 C2J	1 & ½	"	260
64	1 & 1/4 C2JP	1 & ½	"	241
65	1 & 1/2 C1J	2	"	331
66	2 C1J	2	"	353

Jacuzzi:
Max. Temp.—200°F, NPT Conn.

67	3DT1	⅓	3450	165
68	5DT1	½	"	175
69	5DS1	½	"	175
70	5DH1	½	"	175
71	5DH1 & 1/2	½	"	185
72	7DT1	¾	"	185
73	7DS1	¾	"	185
74	7DH1	¾	"	185
75	7DH1 & 1/2	¾	"	195
76	1DS1	1	"	205
77	1DH1	1	"	205
78	1DH1 & 1/2	1	"	215
79	15DH1	1 & ½	"	250
80	15DH1 & 1/2	1 & ½	"	260
81	15DM1	1 & ½	"	275
82	2DH1 & 1/2	2	"	300

Table B-1—Continued

Manufacturer and Other Information	Curve No.	Type	HP	RPM	Cost ($) Iron	Bronze	Stainless Steel
	83	2DM1	2	"	325		
	84	3DM1	3	"	450		
Myers:							
Max. Temp.—200°F Standard,	85	100M	1/3	3450	120		
212°F w/Special Seal for a few	86	100M	1/2	"	125		
dollars more. NPT Conn.	87	100M	3/4	"	140		
	88	QP7	3/4	"	185		
	89	100M	1	"	154		
	90	QP10	1	"	198		
	91	12SM	1 & 1/2	"	225		
	92	QP15	1 & 1/2	"	236		
	93	125M	2	"	265		
	94	QP20	2	"	284		
	95	125M	3	"	307		
	96	QP30	3	"	353		
Oberdorfer:	97	172BUSB	1/8	3450		83	
Max. Temp.—212°F, NPT Conn.	98	300BUSF	1/3	"		155	
Also available in aluminum for	99	500USF	1/3	"		99	
same price.	100	600USF	1/3	"		112	
	101	700AUSF	1/3	"		139	
	102	700BUST	1/2	"		148	
	103	700CUSM	3/4	"		157	
	104	700DUST	1 & 1/2	"		208	
	105	815BUST	1 & 1/2	"		300	
	106	820BUSW	2	"		343	
	107	830BUSY	3	"		403	

Taco:
Max. Temp.—240°F, Flanged Conn.

No.	Model	HP	RPM		
108	007-1	1/25	3450	50	115
109	UN110	1/12	1725	70	160
110	111	1/8	"	106	190
111	113	1/8	"	135	200
112	120	1/6	"	120	312
113	121	1/4	"	172	350
114	122	1/4	"	174	
115	112	1/3	3450	115	170
116	131	1/3	1725	233	400
117	132	1/2	"	245	410
118	133	3/4	"	303	450
119	138	1	"	390	600

Thrush:
Max. Temp.—250°F, Flanged Conn.
Except BC has soldered Conn.

No.	Model	HP	RPM		
120	BCW	1/8	1750		179
121	BC	1/6	"		145
122	HB-1	1/6	"		155
123	1 & 1/2GTA-4	1/4	"	284	
124	1 & 1/2GTBA-4	1/4	"		402
125	1 & 1/2GTA-5 & 1/4	1/3	"	292	
126	1 & 1/2GTBA-5 & 1/4	1/3	"		415

Worthington:
Max. Temp.—225°F, Flanged or
NPT Conn.

No.	Model	HP	RPM		
127	D520 1-1 & 1/4-5	1/3	3500		190

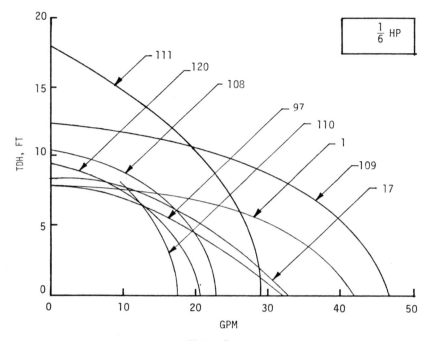

Figure B-1.

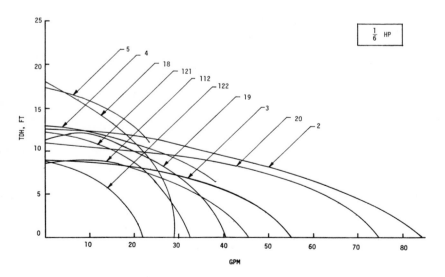

Figure B-2.

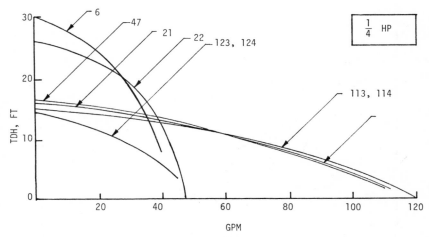

Figure B-3.

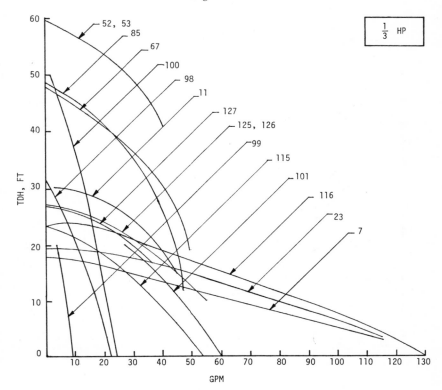

Figure B-4.

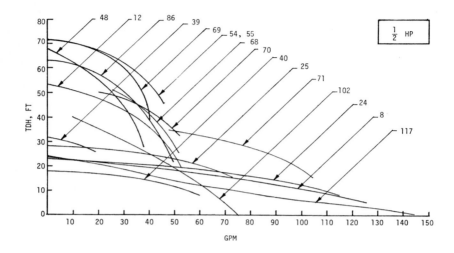

Figure B-5.

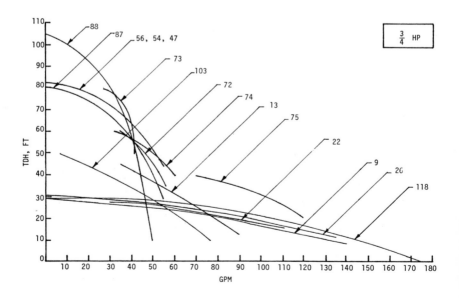

Figure B-6.

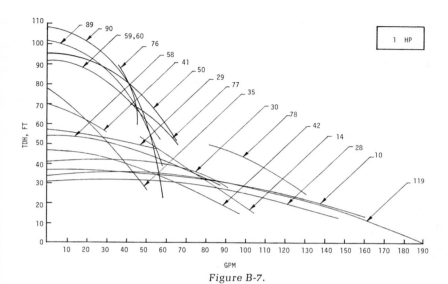

Figure B-7.

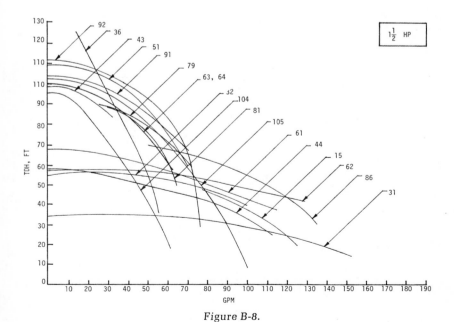

Figure B-8.

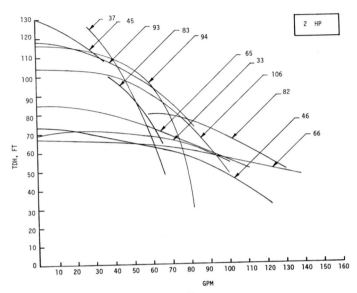

Figure B-9.

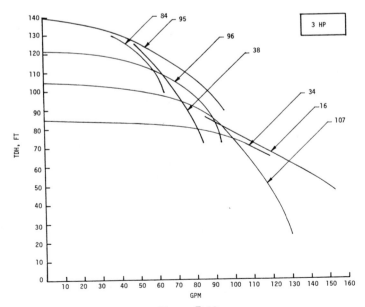

Figure B-10.

Appendix C. Sources of Solar Energy Equipment

Air Heating Solar Collectors

Ametek
Power Systems Group
One Spring Ave.
Hatfield, Pennsylvania 19440
215/248-4600
John C. Bowen

Contemporary Systems in Housing
 Design
Route 1, Box 66
Ashland, New Hampshire 03217
603/968-7841
John C. Christopher

Crimsco, Inc.
5001 East 59th Street
Kansas City, Missouri 64130
816/333-2100
Wayne Nichols

DIY-Sol, Inc.
P.O. Box 614
Marlboro, Maryland 01752
James Barron

Environmental Energies, Inc.
21243 Grand River Avenue
Detroit, Michigan 48219
313/533-1985
Timothy J. Horning

Fun & Frolic, Inc.,
Dept. S
P.O. Box 277
Madison Heights, Michigan 48071
313/399-1560
Edward J. Konopka

Kalwall Corporation
1111 Candia Road
Manchester, New Hampshire
603/627-3861
Drew A. Gillett

Piper Hydro Incorporated
2895 East La Palma
Anaheim, California 92806
714/630-4040
James R. Piper

Solar Energy Products Company
Avon Lake, Ohio 44012
216/933-5000

Solaron Corporation
4850 Olive Street
Denver, Colorado 80022
303/289-2288
Edward M. Redding

Sun-Power Industries
2904 N. Tucson Hiway
Nogales, Arizona 85621
602/287-6620
Al Abramson

Sunwater Company
1112 Pioneer Way
El Cajon, California 92020
Ed Smith

Sunworks, Inc.
669 Boston Post Road
Guilford, Connecticut 06437
203/453-6191

Wormser Scientific Corporation
88 Foxwood Road
Stamford, Connecticut 06903
203/322-1981
Eric M. Wormser

Liquid Heating Solar Collectors

Alpha Designs
Suite 2230
Kroger Building
Cincinnati, Ohio 45202
513/621-1243

Ametek
Power Systems Group
One Spring Ave.
Hatfield, Pennsylvania 19440
215/248-4600

Arizona Solar Enterprises
6718 E. Holly St.
Scottsdale, Arizona 85257
602/945-7477
Arthur M. Walters

Energex Corporation
5115 Industrial Road
Las Vegas, Nevada 89118
702/736-2994

Energy Systems, Inc.
634 Crest Dr.
El Cajon, California 92021
714/440-4646

Timothy J. Horning
21243 Grand River
Detroit, Michigan 48219
313/533-1985

Enviropane, Inc.
348 N. Marshall St.
Lancaster, Pennsylvania 17602
717/299-3737
Joseph M. Bond

Fafco, Inc.
Bohannon Industrial Park
138 Jefferson Drive
Menlo Park, California 94025
415/321-6311
F. Ford Bossche

Falbel Energy Systems Corp.
472 Westover Road
Stamford, Connecticut 06902
203/357-0626
Gerald Falbel

Fun & Frolic, Inc.
Dept. S
P.O. Box 277
Madison Heights, Michigan 48071
313/399-1560
Edward J. Konopka

GB Electronics
1201 Dove St.
Suite 380
Newport Beach, California 92660
714/642-8580
Tom Bernhardt

General Electric Space Division
Valley Forge Space Center
P.O. Box 8555
Philadelphia, Pennsylvania 19101
215/962-4282

Helio-Dynamics, Inc.
518 South Van Ness Avenue
Los Angeles, California 90020
213/384-9853

Independent Living, Inc.
5715 Buford Highway, NE
Doraville, Georgia 30340
404/945-2203
William T. Hudson

International Environment Corp.
1400 Mill Creek Road
Gladwyne, Pennsylvania 19035
215/642-3060
Richard H. Kiley

International Environment Corp.
129 Halstead Avenue
Mamaroneck, New York 10543
914/698-8130
Richard D. Rothschild

International Environmental
   Energy, Inc.
P.O. Box 295
Hartford, Connecticut 06101
203/249-5011
L. Hafetz

InterTechnology Corp.
100 Main St.
Warrenton, Virginia 22186
703/347-7900
David W. Doyle

Mel Kiser & Associates
6701 E. Kenyon Dr.
Tucson, Arizona 85710
602/296-6552
Mel Kiser

KTA Corp.
12300 Washington Ave.
Rockville, Maryland 20852
301/881-0047
Ted Knapp

LO-K Medallion Systems, Inc.
P.O. Box 188
Belmont, North Carolina 28012
825-5357
Robert Kincaid

Martin Marietta Aerospace
Denver Division
P.O. Box 179
Denver, Colorado 80201
303/794-5211

OEM Products, Inc.
Solarmatic Division
214 W. Brandon Blvd.
Brandon, Florida 33511
813/689-1182
D. W. Barlow, Sr.

Owens-Illinois, Inc.
P.O. Box 1035
Toledo, Ohio 43666
419/243-1015
Richard E. Ford

PPG Industries, Inc.
Solar Systems Sales
One Gateway Center
Pittsburgh, Pennsylvania 15222
412/434-3552
Neill M. Barker

Raypak, Inc.
31111 Agoura Rd.
P.O. Box 5790
Westlake Village, California 91359
213/889-1500

Solar Research
Division of Refrigeration
  Research, Inc.
525 N. Fifth St.
Brighton, Michigan 48116
313/227-1151

Revere Copper & Brass, Inc.
P.O. Box 151
Rome, New York 13440
315/338-2295
William J. Heidrich

Reynolds Metals Co.
Mill Products Div.
Richmond, Virginia 23261
804/282-2311
Chester H. Holtyn

Rocky Mountain Solar Heating
  Components
Box 10 Lytle Star Route
Colorado Springs, Colorado 80906
303/576-5266
Dick Gavitt

J. & R. Simmons Construction Co., Inc.
2185 Sherwood Dr.
South Daytona, Florida 32019
904/767-6367
John Simmons

Sol Tex Corporation
P.O. Box 1124
Houston, Texas 77001
713/236-0686
A. E. Cunningham

SOI Solar Development Inc.
4180 Westroads Dr.
West Palm Beach, Florida 33407
305/842-8935
D. Kazimir

Solar Dynamics, Inc.
4527 E. 11th Avenue
Hialeah, Florida 33013
305/688-4393

Solar Energy Modules Company
1091 S. W. 1st Way
Deerfield Beach, Florida 33441
305/427-0040
David B. Aspinwall, Jr.

Solar Energy Systems, Inc.
2492 Banyan Drive
Los Angeles, California 90049
213/472-6508
Kenneth I. Brody

Solar Systems Sales
180 Country Club Dr.
Novato, California 94947
415/883-7040
George Walters

Sol-R-Tech
The Trade Center
Hartford, Vermont 05047
802/295-9342

Solus, Inc.
P.O. Box 35227
Houston, Texas 77035
713/681-1224
Robert R. Barrett

Spectra Energy Systems, Inc.
1210 Camino Rio Verde
Santa Barbara, California 93111
905/964-5584
I. D. Liu

Sunearth, Inc.
P.O. Box 99
Milford Square, Pennsylvania 18935
215/536-8555
Howard Katz

Sun-Power Industries
2904 N. Tuscon Hiway
Nogales, Arizona 85621
602/287-6620
Al Abramson

Sunsav, Inc.
250 Canal St.
Lawrence, Massachusetts 01840
617/686-8040
Peter H. Ottmar

Sunvert Systems, Inc.
Rt. 515, Box 288
Vernon, New Jersey 07462
201/764-4082
Kurt J. Wasserman

Sunwater Company
Solar Energy Products
1112 Pioneer Way
El Cajon, California 92020
Ed Smith

Sunworks, Inc.
669 Boston Post Rd.
Guilford, Connecticut 06437
203/453-6191

Thomason Solar Homes, Inc.
6802 Walker Mill Rd., S.E.
Washington, D.C. 20027
202/336-0009
Harry E. Thomason

U.S. Solar Corp., Inc.
2426 Linden Lane
Silver Springs, Maryland 20901
301/585-2000
B. Kohn

Wallace Sheet Metal Works
831 Dorsey St.
P.O. Box 511
Gainesville, Georgia 30501

Western Energy
454 Forest Avenue
Palo Alto, California 94302

Wilson Solar Kinetics Corp.
P.O. Box 876
Hartford, Connecticut 06101
203/633-7884
James A. Pholman

Wormser Scientific Corp.
88 Foxwood Rd.
Stamford, Connecticut 06903
203/322-1981
Eric M. Wormser

Control Units for Solar System

DIY-Sol., Inc.
P.O. Box 614
Marlboro, Maryland 01752
Felix Rapp

Electro-Logic Fabricators
P.O. Box 1068
State College, Pennsylvania 16801
Bill Harding

Fafco, Inc.
Bohannon Industrial Park
138 Jefferson Dr.
Menlo Park, California 94025
415/321-6311
F. Ford VandenBossche

Deko-Labs
P.O. Box 12841
University Station
3860 S.W. Archer Rd., H3
Gainesville, Florida 32604
904/372-6009
Donald F. DeKold

General Electric Space Division
Valley Forge Space Center
P.O. Box 8555
Philadelphia, Pennsylvania 19101
215/962-4282

Honeywell
Honeywell Plaza
Minneapolis, Minnesota 55408
612/870-2278
Robert W. Stalder

OEM Products, Inc.
Solarmatic Division
214 W. Brandon Blvd.
Brandon, Florida 33511
813/689-1182
D. W. Barlow, Sr.

RANCO
601 W. Fifth Ave.
P.O. Box 8187
Columbus, Ohio 43201
614/294-3511
R. G. Raney

Rho Sigma
5108 Melvin Avenue
Tarzana, California 91356
213/342-4376
R. J. Schlesinger

Solar Dynamics Inc.
4527 E. 11th Avenue
Hialeah, Florida 33012
305/688-4393

SEMCO
1091 S.W. 1st Way
Deerfield Beach, Florida 33441
305/427-0040

Solar Sensor System
4220 Berritt St.
Fairfax, Virginia 22030
703/273-2683
Jack S. Scovel

Solar-Temp
725 Federal Avenue
Kenilworth, New Jersey 07033
201/245-3190
Richard N. Foster

Trol-A-Temp
725 Federal Ave.
Kenilworth, New Jersey 07033
201/245-3190
Richard N. Foster

VAST, Inc.
Product Development Division
330 Boston Post Road
P.O. Box 665
Old Saybrook, Connecticut 06475
203/388-3429
J. Carnell

VERTREX Corp.
208 Carlson Bldg.
808 106th N.E.
Bellevue, Washington 98004
206/455-4718
Albert H. Rooks

Zia Associates, Inc.
5590 Arapahoe
P.O. Box 1466
Boulder, Colorado 80302
303/449-9170
Thomas B. Kent

Sunwater Co.
Solar Energy Products
1112 Pioneer Way
El Cajon, California 92020
Ed Smith

Solar Storage Tanks

Owens-Corning Fiberglas Corp.
Non-Corrosive Products Division
Fiberglas Tower
Toledo, Ohio 43659
419/259-2120
Q. V. Meeks

Solar Central
7213 Ridge Road
Mechanicsburg, Ohio 43044
513/322-2533
Donald Greider

Solar Stills

Sunwater Company
1112 Pioneer Way
El Cajon, California 92020
Ed Smith

Solar Air Conditioning Units

Solaire
Arkla Industries, Inc.
Special Products Division
400 East Capitol
Little Rock, Arkansas 72203
501/372-6241
Thomas M. Helms

Barber-Nichols Engineering Co.
6325 W. 55th Avenue
Arvada, Colorado 80002
303/421-8111
Robert E. Barber

York
York Division of Borg-Warner Corp.
P.O. Box 1592
York, Pennsylvania 17405

Chrysler Corp.
P.O. Box 7806
Cape Canaveral, Florida 32920
305/783-5900
T. F. McCrea

Solar Electric Power Generation

Barber-Nichols Engineering Co.
6325 W. 55th Ave.
Arvada, Colorado 80002
303/421-8111
Robert E. Barber

Bay View Electric Co.
Solar Research Division
P.O. Box 4453
Bay View, Wisconsin 53207
James Kilowatt

Solar Energy Co.
818 18th St., N.W.
Washington, D.C. 20006
G. H. Hamilton

Solar Intensity Instruments

The Eppley Laboratory, Inc.
12 Sheffield Ave.
Newport, Rhode Island 02840
401/847-1020
George L. Kirk

International Scientific Industries
P.O. Box 537
Flagstaff, Arizona
602/774-8926
Herbert Wade

IBM
Federal Systems Division
150 Sparkman Drive
Huntsville, Alabama 35805
Ms. P. A. Cosper

Molectron Corp.
177 N. Wolfe Rd.
Sunnyvale, California 94086
408/738-2661
George Lee

Solar Research
Solar Thermal Division
17050 Chatsworth
Suite 242
Granada Hills, California 91344
213/368-1311
Lothar R. Wollmann

Sun Research
P.O. Box 16166
Jacksonville, Florida 32216
904/641-2484

Suntek Research Associates, Inc.
33 Edinboro St.
Boston, Massachusetts
617/482-1245
John Brookes

REFERENCES

1. "Reference Energy Systems and Resource Data", *Associated Universities, Inc.,* AET-8, April 1972.
2. Cherry, William R., "Harnessing Solar Energy: The Potential", *Astronautics & Aeronautics,* p. 30–36, Aug. 1973.
3. Rose, David J., "Energy Policy in the U.S.", *Scientific American,* Vol. 230, No. 1, pp. 20–29, January 1974.
4. Gaucher, Leon P., "The Solar Era: Part I—The Practical Promise", *Mechanical Engineering,* pp. 9–12, August 1972.
5. Gambs, Gerard C., *The Energy Crisis in the United States,* published by Ford, Bacon & David, Inc., March 1973.
6. Ritchings, Frank A., "Trends in Energy Needs", *Mechanical Engineering,* pp. 18–23, August 1972.
7. Nixon, Richard M., "Energy Policy", Message to the Congress Announcing Executive Actions and Proposing Enactment of Bills to Provide for Energy Needs, *Presidential Documents,* Vol. 9, No. 16, pp. 389–406, April 18, 1973.
8. Dupree, W. G., Jr. and West, James A., "United States Energy Through the Year 2000", *Dept. of Interior,* U.S. Govt. Printing Office, 1973.
9. Schulman, Fred, "Technology, The Energy Crisis, and our Standard of Living", *Mechanical Engineering,* pp. 16–23, September 1973.
10. "Toward a National Energy Policy", *Environmental Science & Technology,* Vol. 7, No. 5, p. 392–397, May 1973.
11. Altman, M., Telkes, M. and Wolf, M., "The Energy Resources and Electric Power Situation in the United States", *Energy Conversion,* Vol. 12, pp. 63–64, 1972.
12. Gambs, G. C. and Rauth, A. A., "The Energy Crisis", *Chemical Engineering,* p. 56–68, May 31, 1971.
13. Donovan, Paul and Woodward, William, "An Assessment of Solar Energy As a National Energy Resource", *NSF/NASA Solar Energy Panel* (Univ. of Maryland), December 1972.

14. "Patterns of Energy Consumption in the United States", *Stanford Research Institute Report*, January 1972.
15. Löf, G. O. G. and Tybout, R. A., "Cost of House Heating with Solar Energy", *Solar Energy*, Vol. 14, pp. 253–278, 1973.
16. "Energy Research and Development Space Technology", *Hearings of the Committee on Science and Astronautics*, U.S. House of Representatives, U.S. Government Printing Office, Washington, D.C., May 1973.
17. Böer, K. W., "Future Large Scale Terrestrial Use of Solar Energy", *Proceedings of the NASA Conference on Solar and Chemical Power*, QC-603, p. 145–148, 1972.
18. Löf, G. O. G., Duffie, J. A. and Smith, C. O., "World Distribution of Solar Energy", *Solar Energy*, Vol. 10, No. 1, pp. 27–37, 1966.
19. *Transactions of the NSF/NOAA Solar Energy Data Workshop*, Washington, D.C., Nov. 29–30, 1973, In press.
20. Lorsch, Harold G., "Performance of Flat Plate Collectors", *Proceedings of the NSF/RANN Solar Heating and Cooling for Buildings Workshop*, Washington, D.C., NSF/RANN–73–004, pp. 1–14, July 1973.
21. Streed, Elmer R., "Some Design Considerations for Flat Plate Collectors", *Proceedings of the NSF/RANN Solar Heating and Cooling for Buildings Workshop*, NSF/RANN–73–004, pp. 26–35, July 1973.
22. Merriam, Marshall, E., "Materials Technology for Flat Plate Steam Generation", *Proceedings of the NSF/RANN Solar Heating and Cooling for Buildings Workshop*, NSF/RANN–73–004, pp. 17–20, July 1973.
23. Böer, Karl W., "Solar Collectors of the University of Delaware Solar House Project", *Proceedings of the NSF/RANN Solar Heating and Cooling for Buildings Workshop*, NSF/RANN–73–004, pp. 15–16, July 1973.
24. Edmondson, W. B., ed., "Breakthrough in Selective Coatings", *Solar Energy Digest*, Vol. 2, No. 2, p. 1–2, February 1974.
25. Thomason, H. E., "Solar Houses and Solar House Models", Edmund Scientific Company, Barrington, N. J., 1972.
26. Thomason, H. E. and Thomason, H. J. L., Jr., "Solar Houses/Heating and Cooling Progress Report", *Solar Energy*, Vol. 15, pp. 27–39, 1973.
27. NASA, "The Development of a Solar-Powered Residential Heating and Cooling System," NASA Report M-TU-74-3, May 1974.
28. Moore, S. W., Balcomb, J.D. and Hedstrom, J. C., "Design and Testing of a Structurally Integrated Steel Solar Collector Unit Based on Expanded Flat Metal Plates," LASL Report LA-UR-74-1093, presented at the U.S. Sections of the ISES Annual Meeting, Fort Collins, Colorado, August 1974.
29. McDonald, G. E., "Spectral Reflectance Properties of Black Chrome

30. Teplyakov, D. I., "Analytic Determination of the Optical Characteristics of Paraboloidal Solar Energy Concentrators," *Geliotekhnika*, Vol. 7, No. 5, pp. 21–33, 1971 (Russian).

31. Giutronich, J. E., "The Design of Solar Concentrators Using Toroidal Spherical, or Flat Components", *Solar Energy*, Vol. 7, No. 4, pp. 162–166, 1963.

32. Eibling, James A., "A Survey of Solar Collectors", *Proceedings of the NSF/RANN Solar Heating and Cooling for Buildings Workshop*, NSF/RANN–73–004, pp. 47–51, July 1973.

33. Lidorenko, N. S., Nabiullin, F. Ka, Landsman, A. P., Tarnizhevskii, B. V., Gertsik, E. M. and Shul'meister, L. F., "An Experimental Solar Power Plant", *Geliotekhnika*, Vol. 1, No. 3, pp. 5–9, 1965 (Russian).

34. Gunter, Carl, "The Utilization of Solar Heat for Industrial Purposes by Means of a New Plane Mirror Reflector", *Scientific American*, May 26, 1906.

35. Steward, W. Gene, "A Concentrating Solar Energy System Employing a Stationary Spherical Mirror and Movable Collector", *Proceedings of the NSF/RANN Solar Heating and Cooling for Buildings Workshop*, NSF/RANN–73–004, pp. 17–20, July 1973.

36. Russell, J. L., DePlomb, E. P. and Ravinder, R. K., "Principles of the Fixed Mirror Solar Concentrator", *Solar Energy*, submitted for publication.

37. "Solar Energy Research—A Multi-disciplinary Approach", *Staff Report of the Committee on Science and Astronautics of the House of Representatives*, December 1972.

38. Weingart, J. M. and Schoen, Richard, "Project SAGE—An Attempt to Catalyze Commercialization of Gas Supplemented Solar Water Heating Systems for New Apartments in Southern California", *Proceedings of the NSF/RANN Solar Heating and Cooling for Buildings Workshop*, NSF/RANN–73–004, pp. 75–91, July 1973.

39. Lorsch, Harold G., "Solar Heating/Cooling Projects at the University of Pennsylvania", *Proceedings of the NSF/RANN Solar Heating and Cooling for Buildings Workshop*, NSF/RANN–73–004, pp. 194–210, July 1973.

40. Close, D. J., "Solar Air Heaters", *Solar Energy*, Vol. 7, pp. 117–124, 1963.

41. Satcunanthan, Sand Deonarine, S., "A Two-Pass Solar Air Heater", *Solar Energy*, Vol. 15, pp. 41–49, 1973.

42. Telkes, Maria, "Energy Storage Media", *Proceedings of the NSF/RANN Solar Heating and Cooling for Buildings Workshop*, NSF/RANN–73–004, pp. 57–59, July 1973.

43. Hay, H. R., and Yellott, J. I., "A Naturally Air-Conditioned Building", *Mechanical Engineering*, Vol. 92, No. 1, pp. 19–25, Jan. 1970.

44. Hay, H. R., "Energy Technology and Solarchitecture", *Mechanical Engineering*, pp. 18–22, Nov. 1973.

45. Hay, Harold R., "Evaluation of Proved Natural Radiation Flux Heating and Cooling", *Proceedings of the NSF/RANN Solar Heating and Cooling for Buildings* Workshop, NSF/RANN–73–004, pp. 185–187, July 1973.

46. Tybout, R. A. and Löf, G. O. G., "Solar House Heating", *Natural Resources Journal*, Vol. 10, No. 2, p. 268–326, April 1970.

47. Noguchi, Tetsuo, "Recent Developments in Solar Energy Research Application in Japan", *Solar Energy*, Vol. 15, pp. 179–187, 1973.

48. Sobotka, R., "Economic Aspects of Commercially Produced Solar Water Heaters", *Solar Energy*, Vol. 10, No. 1, pp. 9–14, 1966.

49. Malik, M. A. S., "Solar Water Heating in South Africa", *Solar Energy*, Vol. 12, p. 395–397, 1969.

50. Daniels, Farrington, "Direct Use of the Sun's Energy", *American Scientist*, Vol. 55, No. 1, pp. 29–30, 1967.

51. Farben, E. A., Flanigan, F. M., Lopez, L. and Politka, R. W., "University of Florida Solar Air-Conditioning System", *Solar Energy*, Vol. 10, pp. 91–99, 1966.

52. Teagen, W. P., "A Solar Powered Heating/Cooling System with the Air Conditioning Unit Driven by an Organic Rankine Cycle Engine", *Proceedings of the NSF/RANN Solar Heating and Cooling for Buildings Workshop*, NSF/RANN 73–004, pp. 107–111, July 1973.

53. Baum, V. A., Aparase, A. R. and Garf, B. A., "High-Power Solar Installations", *Solar Energy*, Vol. 1, No. 2, pp. 6–13, 1957.

54. Edlin, F. E., "Worldwide Progress in Solar Energy", *Proceedings of the Intersociety Energy Conversion Engineering Conference* (IECEC), pp. 92–97, New York, 1968.

55. Hildebrandt, A. F. and Vant-Hull, L. L., "Large Scale Utilization of Solar Energy", *Energy Research and Development,* Hearings of the House Committee on Science and Astronautics, pp. 499–505, U.S. Govt. Printing Office, 1972.

56. Trombe, F., "Solar Furnaces and Their Applications", *Solar Energy,* Vol. 1, No. 2, pp. 9–15, 1957.

57. Walton, J. D., personal communication.

58. Meinel, A. B. and Meinel, M. P., "Energy Research and Development", *Hearings of the House Committee on Science and Astronautics,* U.S. Govt. Printing Office, pp. 583–585, 1972.

59. Russell, John L., Jr., "Investigation of a Central Station Solar Power Plant", *Proceedings of the Solar Thermal Conversion Workshop,* Washington, D.C., January 1973; (also published as *Gulf General Atomic Report* No. Gulf-GA-A12759, August 31, 1973).

60. Oman, H. and Bishop, C. J., "A Look at Solar Power for Seattle", *Proc. IECEC,* pp. 360–65, August 1973.

61. Meinel, A. B., "A Joint United States-Mexico Solar Power and Water

Facility Project", Optical Sciences Center, University of Arizona, April 1971.

62. Ralph, E. L., "A Commercial Solar Cell Array Design", *Solar Energy*, Vol. 14, pp. 279–286, 1973.

63. Currin, C. G., Ling, K. S., Ralph, E. L., Smith, W. A. and Stirn, R. J., "Feasibility of Low Cost Silicon Solar Cells", *Proceedings of the 9th IEEE Photovoltaic Specialists Conference*, Silver Spring, Md., May 2–4, 1972.

64. Riel, R. K., "Large Area Solar Cells Prepared on Silicon Sheet", *Proceedings of the 17th Annual Power Sources Conference*, Atlantic City, N.J., May 1963.

65. Cherry, W. R., *Proceedings 13th Annual Power Sources Conference*, Atlantic City, N.J., pp. 62–66, May 1959.

66. Tyco Laboratories, *Final Report* #AFCRL–66–134, 1965.

67. Currin, C. G., Ling, K. S., Ralph, E. L., Smith, W. A. and Stirn, R. J., "Feasibility of Low Cost Silicon Solar Cells", *Proceedings of the 9th IEEE Photovoltaic Specialists Conference*, Silver Spring, Md., May 2–4, 1972.

68. Eckert, J. A., Kelley, B. P., Willis, R. W. and Berman, E., "Direct Conversion of Solar Energy on Earth, Now", *Proc. IECEC*, pp. 372–5, 1973.

69. Homer, Lloyd, *NSF/NOAA Solar Energy Data Workshop*, Washington, D.C., Nov. 1973.

70. Rink, J. E. and Hewitt, J. G., Jr., "Large Terrestrial Solar Arrays", *Proceedings of the Intersociety Energy Conversion Engineering Conference*, Boston, Mass., Aug. 3–5, 1971.

71. Wolf, M., *Proceedings of the 9th IEEE Photovoltaic Specialists Conference*, Silver Spring, Md., May, 1972.

72. Ralph, E. L., "A Plan to Utilize Solar Energy as an Electric Power Source", *IEEE Photovoltaic Specialists Conference*, pp. 326–330, 1970.

73. Pfeiffer, C., Schoffer, P., Spars, B. G. and Duffie, J. A., "Performance of Silicon Solar Cells at High Levels of Solar Radiation", *Transactions of the American Society of Mechanical Engineers (ASME)*, 84A, p. 33, 1962.

74. Pfeiffer, C. and Schoffer, P., "Performance of Photovoltaic Cells at High Radiation Levels", *Trans. ASME*, 85A, p. 208, 1963.

75. Beckman, W. A., Schoffer, P., Hartman, W. R., Jr. and Löf, G. O. G., "Design Consideration for a 50 Watt Photovoltaic Power System Using Concentrated Solar Energy", *Solar Energy*, Vol. 10, No. 3, pp. 132–136, 1966.

76. Tarnizhevskir, B. V., Savchenko, I. G., and Rodichev, B. Y., "Results of an Investigation of a Solar Battery Power Plant", *Geliotekhnika*, Vol. 2, No. 2, pp. 25–30, 1966 (Russian).

77. "Solar Energy", *Moscow News,* No. 2, January 1974.

78. Farber, E. A., "Solar Energy, Its Conversion and Utilization", *Solar Energy,* Vol. 14, pp. 243–252, 1973.

79. Schaeper, H. R. A. and Farber, E. A., "The Solar Era: Part 4—The University of Florida Electric", *Mechanical Engineering,* pp. 18–24, November 1972.

80. Backus, C. E., "A Solar-Electric Residential Power System", *Proc. IECEC,* 1972.

81. Rabenhorst, D. W., "Superflywheel", *Proceedings of the NSF/RANN Solar Heating and Cooling for Buildings Workshop,* NSF–RANN–73–004, pp. 60–68, July 1973.

82. Loferski, J. J., "Large Scale Solar Power Via the Photoelectric Effect", *Mechanical Engineering,* December, 1973, pp. 28–32.

83. Pope, R. B., *et al.,* "A Combination of Solar Energy and the Total Energy Concept—The Solar Community", *Proc. 8th IECEC,* pp. 304–311, 1973.

84. Schimmel, W. P., Jr., "A Vector Analysis of the Solar Energy Reflection Process", *Annual Meeting of the U.S. Section of the International Solar Energy Society,* Cleveland, Ohio, October 1973.

85. Pope, R. B. and Schimmel, W. P., Jr., "An Analysis of Linear Focused Collectors for Solar Power", *Proc. 8th IECEC,* Philadelphia, Penn., Aug. 1973.

86. Achenbach, P. R. and Cable, J. B., "Site Analysis for the Applications of Total Energy Systems to Housing Developments", *7th IECEC,* San Diego, Ca., Sept. 1972.

87. Sakvrai, T., Osamu, K., Koro, S. and Koji, I., "Construction of a Large Solar Furnace", *Solar Energy,* Vol. 8, No. 4, 1964.

88. Akyurt, M. and Selcuk, M. K., "A Solar Drier Supplemented with Auxiliary Heating Systems for Continuous Operation", *Solar Energy,* Vol. 14, pp. 313–320, 1973.

89. Bevill, V. D. and Brandt, H., "A Solar Energy Collector for Heating Air", *Solar Energy,* Vol. 12, pp. 19–29, 1968.

90. Khanna, M. L., "Design Data for Heating Air by Means of Heat Exchanger with Reservior, Under Free Convection Conditions, For Utilization of Solar Energy", *Solar Energy,* Vol. 12, pp. 447–456, 1969.

91. Hay, Harold R., "The Solar Era: Part 3—Solar Radiation: Some Implications and Adaptations", *Mechanical Engineering,* pp. 24–29, October 1972.

92. Morse, R. N. and Read, W. R. W., "A Rational Basis for the Engineering Development of a Solar Still", *Solar Energy,* Vol. 12, pp. 5–17, 1968.

93. Lewand, T. A., "Description of a Large Solar Distillation Plant in the West Indies", *Solar Energy,* Vol. 12, pp. 509–512, 1969.

94. Hay, Harold R., "New-Roofs for Hot Dry Regions", *Ekistics*, Vol. 31, pp. 158–164, 1971.

95. Talbert, S. G., *et al.*, Manual on Solar Distillation of Saline Water, *Progress Report No. 546*, U.S. Dept. of Interior, Office of Saline Water, 1970.

96. Achilov, B., *et al.*, "Investigation of an Industrial-Type Solar Still", *Geliotekhnika*, Vol. 7, No. 2, pp. 33–36, 1971 (Russian).

97. Achilov, B. M., "Comparative Tests on Large Solar Stills in the Fields of Kzyl-kum in the Uzbek SSR", *Geliotekhnika*, Vol. 7, No. 5, pp. 86–89, 1971 (Russian).

98. Baum, V. A. and Bairamov, R., "Heat and Mass Transfer Processes in Solar Stills of the Hotbox Type", *Solar Energy*, Vol. 8, No. 3, pp. 78–82, 1964.

99. Annaev, A., Bairamov, R., and Rybakova, L. E., "Effect of Wind Speed and Direction on the Output of a Solar Still", *Geliotekhnika*, Vol. 7, No. 4, pp. 33–37, 1971 (Russian).

100. Szego, G. C., Fox, J. A., and Eaton, D. R., "The Energy Plantation", Paper no. 729168, *Proc. IECEC*, pp. 113–4, September 1972.

101. Woodmill, G. W., "The Energy Cycle of the Biosphere", *Scientific American*, Vol. 233, No. 3, p. 70, September 1970.

102. DeBeni, G. and Marchetti, C., "Hydrogen, Key to the Energy Market", *Eurospectra*, Vol. 9, No. 2, p. 46, 1970.

103. Marchetti, C., "Hydrogen and Energy", *Chemical Economy and Engineering Review*, pp. 7–25, January, 1973.

104. D'Arsonval, J., *Revue Scientifique*, Vol. 17, September 1881.

105. Claude, Georges, "Power From the Tropical Seas", *Mechanical Engineering*, Vol. 52, No. 12, pp. 1039–1044, December 1930.

106. Roe, Ralph C. and Othmer, Donald F., "Controlled Flash Evaporation", *Mechanical Engineering*, Vol. 93, No. 5, pp. 27–31, 1971.

107. *Proceedings of the Solar Sea Power Plant Conference and Workshop*, sponsored by the National Science Foundation (RANN), Carnegie-Mellon Univ., Pittsburgh, Pa., June, 1973.

108. Anderson, J. H. and Anderson, J. H., Jr., "Thermal Power from Seawater", *Mechanical Engineering*, Vol. 88, No. 4, pp. 41–46, April 1966.

109. Zener, Clarence, "Solar Sea Power", *Physics Today*, pp. 48–53, January 1973.

110. McGowan, J. G., Connell, J. W., Ambs, L. L. and Goss, W. P., "Conceptual Design of a Rankine Cycle Powered by the Ocean Thermal Difference", *Proc. IECEC*, paper no. 739120, pp. 420–27, August 1973.

111. Anderson, J. H., "The Sea Plant—A Source of Power, Water and Food Without Pollution", *Solar Energy*, Vol. 14, pp. 287–300, 1973.

112. Barnes, S., "Mining Marine Minerals", *Machine Design*, April, 1968.

113. Banks, Ridgway, personal communication, January 1974.

114. a) Glaser, P. E., "The Future of Power from the Sun", *Proc. IECEC*,

pp. 98–103, 1968, and b) "Power From the Sun: Its Future", *Science,* Vol. 162, pp. 857–861, Nov. 22, 1968.

115. Glaser, P. E., "A New View of Solar Energy", *Proc. IECEC,* paper no. 719002, pp. 1–4, 1971.

116. Glaser, P. E., Maynard, O. E., Mockovciak, J. and Ralph, E. L. "Feasibility Study of a Satellite Solar Power Station", *NASA Contractor Report* CR-2357, February 1974.

117. Glaser, Peter E., "Solar Power Via Satellite", *Astronautics and Aeronautics,* pp. 60–68, August 1973.

118. Brown, William C., "Adapting Microwave Techniques to Help Solve Future Energy Problems", *Proceedings of the IEEE International Microwave Symposium,* pp. 189–191, June 1973.

119. Patha, J. T. and Woodcock, G. R., "Feasibility of Large-Scale Orbital Solar/Thermal Power Generation", *Proc. IECEC,* pp. 312–319, 1973.

120. Brown, William C., "Satellite Power Stations: A New Source of Energy?", *IEEE Spectrum,* pp. 38–47, March 1973.

121. Mockovciak, John, Jr., "A Systems Engineering Overview of the Satellite Power Station", *Proc. 7th IECEC,* paper no. 739111, pp. 712–19, Sept. 1972.

122. Bergey, Karl H., "Wind Power Demonstration and Siting Problems." *Wind Energy Conversion Systems Workshop Proceedings,* NSF/RA/W–73–006, ed. by J. M. Savino, pp. 41–45, December 1973.

123. Wilcox, Carl, "Motion Picture History of the Erection and Operation of the Smith-Putnam Wind Generator", *Wind Energy Conversion Systems Workshop,* NSF/RA/W–73–006, ed. by J. M. Savino, pp. 8–10, December 1973.

124. Smith, B. E., "Smith-Putnam Wind Turbine Experiment", *Wind Energy Conversion Systems Workshop Proceedings,* NSF/RA/W–73–006, ed. by J. M. Savino, pp. 5–7, December 1973.

125. Thomas, Percy H., "Electric Power from the Wind", Federal Power Commission, Washington, D.C., March 1945.

126. Hutter, Ulrich, "Past Developments of Large Wind Generators in Europe", *Wind Energy Conversion Systems Workshop Proceedings,* NSF/RA/W–73–006, pp. 19–22, December 1973.

127. Tompkin, J., "Introduction to Voigt's Wind Power Plant", *Wind Energy Conversion Systems Workshop Proceedings,* NSF/RA/W–73–006, ed. by J. M. Savino, pp. 23–26, December 1973.

128. Clews, Henry M., "Wind Power Systems for Individual Applications", *Wind Energy Conversion Systems Workshop Proceedings,* NSF/RA/W–73–006, pp. 164–69, December 1973.

129. Noel, John M., "French Wind Generator Systems", *Wind Energy Conversion System Workshop Proceedings,* NSF/RA/W–73–006, pp. 186–196, December 1973.

130. Edmondson, William B., "A 100 KW Windmill Generator", *Solar Energy Digest,* Vol. 2, No. 4, p. 4, April 1974.

GLOSSARY

Absorption cooling. Refrigeration or air conditioning achieved by an absorption-desorption process that can utilize solar heat to produce a cooling effect.

Absorptivity. The ratio of the incident radiant energy absorbed by a surface to the total radiant energy falling on the surface.

Albedo. The ratio of the light reflected by a surface to the light falling on it.

Ambient temperature. Prevailing temperature outside a building.

Anaerobic fermentation. Fermentation process caused by bacteria in the absence of oxygen.

Bio-conversion. Use of sunlight to grow plants with subsequent use of the plants to provide energy.

Brayton cycle. Power plant using a gas turbine to drive a compressor and produce power. A gas is compressed, then heated, then expanded through a turbine, then cooled. The turbine produces more power than is needed to drive the compressor.

British Thermal Unit (BTU). A unit of energy which is equal to the amount of heat required to raise the temperature of a pound of water one degree Fahrenheit.

Capital cost. The cost of construction, including design costs, land costs, and other costs necessary to build a facility. Does not include operating costs.

Capture efficiency of plants. The ratio of the energy absorbed and converted into tissue by plants to the total solar energy falling on the plants. This energy, usually about 3% or less of the total incident solar energy, can be released when the plants are burned.

Collector efficiency. The ratio of the energy collected by a solar collector to the radiant energy incident on the collector.

Concentration ratio (concentration factor). Ratio of radiant energy intensity at the hot spot of a focusing collector to the intensity of unconcentrated direct sunshine at the collector site.

Convective heat transfer. Transfer of heat by the circulation of a liquid or gas.

Degree day (DD). One day with the average ambient temperature one degree colder than 65°F. For example, if the average temperature is 55°F for 3 days, the number of degree days is (65–55) times 3, or 30.

Diffuse insolation. Sunlight scattered by atmospheric particulates that arrives from a direction other than the direction of direct sunlight. The blue color of the sky is an example of diffuse solar radiation.

Direct conversion. Conversion of sunlight directly into electric power, instead of collecting sunlight as heat and using the heat to produce power. Solar cells are direct conversion devices.

Direct insolation. Sunlight arriving at a location that has not been scattered, also referred to as direct beam radiation.

Dynamic conversion. The collection of sunlight to heat a fluid that operates an engine to produce power. An example is the use of concentrated sunlight to boil water and operate a steam engine to produce power.

Electrolysis. The use of an electric current to produce hydrogen and oxygen from water.

Equilibrium temperature. The temperature of a device or fluid under steady operating conditions.

Faceted concentrator. A focusing collector using many flat reflecting elements to concentrate sunlight at a point or along a line.

Fossil fuel. Coal, oil or natural gas.

Fuel cell. A device, somewhat like a battery, that uses a chemical reaction to produce electricity directly, such as the reaction of hydrogen and oxygen to produce electric power with water as a product.

Geosynchronous satellite (synchronous satellite). An artificial satellite in a synchronous orbit 22,300 miles from the earth that can remain continuously above the same spot on the earth, since the period of the orbit is 24 hours.

Heat of fusion. The heat released when a liquid becomes a solid (freezes), equal to the heat absorbed when a solid melts—often tabulated in units of BTU/pound.

Heat transfer fluid. A liquid or gas that transfers heat from a solar collector to its point of use.

Heliostat. An electro-optical-mechanical device that orients a mirror so that sunlight is reflected from the mirror in a fixed specific direction, regardless of the sun's position in the sky.

Hot spot. The location on a focusing collector at which the concentrated sunlight is focused and the highest temperatures are produced. If the heat is to be collected, a heat exchanger is located at the hot spot and a heat transfer fluid flowing through the heat exchanger is heated.

Incidence angle. The angle between the direction of the sun and the perpendicular to the surface on which sunlight is falling.

Infrared radiation. Thermal radiation or light with wavelengths longer than 0.7 microns. Invisible to the naked eye, the heat radiated by objects at less than 1000°F is almost entirely infrared radiation.

Insolation. Sunlight, or solar radiation, including ultraviolet, visible and infrared radiation from the sun. Total insolation includes both direct and diffuse insolation.

KWe. Kilowatt of electric power.

KWH. Kilowatt-hour.

KWt. Kilowatt of thermal (heat) energy.

Linear concentrator. A solar concentrator which focuses sunlight along a line, such as the parabolic trough concentrator (Figure 10) and the fixed-mirror concentrator (Figure 30).

MBTU. Million BTU's.

Micron. A millionth of a meter, or micro-meter, a common unit for measuring the wavelength of light. Ultraviolet light has wavelengths less than 0.4 microns, visible light covers the wavelength range of 0.4 to 0.7 microns, and infrared radiation has wavelengths longer than 0.7 microns.

Microscale data. Refers to data on insolation and weather parameters that can vary considerably over distances of a few miles, for example there may be more cloudiness and haze near a lake or in a city than a few miles away. Microscale data can be collected by satellites.

MWe. Megawatt (million watts) of electric power.

MWt. Megawatt of thermal (heat) energy.

Mill. An amount of money equal to one-tenth of a cent.

Optical coatings. Very thin coatings applied to glass or other transparent materials to increase the transmission (reduce the reflection) of sunlight. Coatings are also used to reflect back to the heat exchanger infrared radiation emitted from it.

Phase-change material. A material used to store heat by melting. Heat is later released for use as the material solidifies.

Photovoltaic cells (solar cells). Semiconducting devices that convert sunlight directly into electric power. The conversion process is called the *photovoltaic effect.*

Pyranometer. An instrument for measuring sunlight intensity. It usually measures total (direct plus diffuse) insolation over a broad wavelength range.

Pyrheliometer. An instrument that measures the intensity of the direct beam radiation (direct insolation) from the sun. The diffuse component is not measured.

Radian. A unit of angular measure, one radian equals 57.296 degrees. The sunshine has an angular diameter of 0.009 radians, or one-half degree.

Reflectivity (reflectance). The ratio of light reflected from a surface to the light falling on the surface. The reflectivity plus the absorptivity equals one, since the incident sunlight is either reflected or absorbed.

Selective coating. An optical coating for heat exchangers that has a high absorptivity (low reflectivity) for incident sunlight (wavelengths less than one micron) and high reflectivity (low absorptivity) for infrared heat (wavelengths greater than one micron), as shown in Figure 2. The low infrared absorptivity (low emissivity) results in reduced radiant heat loss, so the collection efficiency is improved, and higher temperatures can be achieved.

Solar concentrator. Device using lenses or reflecting surfaces to concentrate sunlight.

Solar farm. A large array of solar collectors, as shown in Figure 30, for generating large amounts of electric power.

Solar furnaces. Solar concentrators for producing very high temperatures. Installations in France, Russia and Japan produce temperatures as high as 7000°F.

Solar-thermal conversion. The collection of sunlight as heat, and the conversion of heat into electric power.

Specific heat. The amount of heat required to raise the temperature of one pound of material one degree Fahrenheit, usually measured in BTU/lb°F.

Spectral pyranometer. An instrument for measuring total insolation over a restricted wavelength range.

Specular reflection. Mirror-like reflection from a surface.

Thermal efficiency. The ratio of electric power produced by a power plant to the amount of heat supplied to the plant.

Thermochemical hydrogen production. Use of heat with a series of chemical reactions to produce hydrogen from water.

Total energy system. System for providing all energy requirements, including heat, air conditioning, and electric power.

Turbidity. Atmospheric haze.

Refractory materials. Materials that can withstand high temperatures without melting.

Vapor cycle. Method of converting heat into power by boiling a liquid, expanding the vapor through a turbine, condensing the vapor back to a liquid and pumping the liquid back to the boiler. The power output of the turbine is much greater than the power required by the pump.

Waste heat. Heat rejected by a power plant.

INDEX

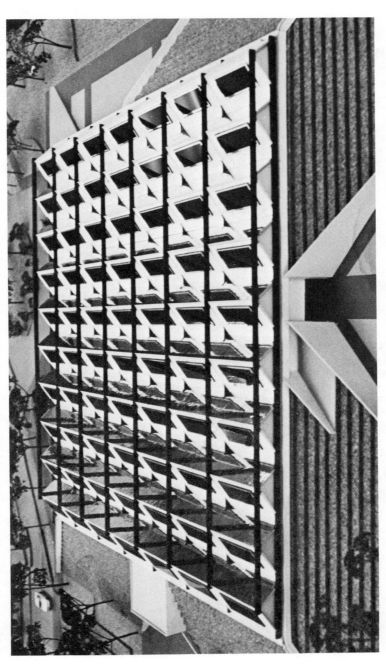

Shenandoah Solar Community Center in Shenandoah, Georgia, south of Atlanta. Sixty-three solar collector modules, each 8 × 21 ft, provide 90% of the heating, 60% of the cooling, domestic hot water and swimming pool heating.

University of Florida Solar House (1956) heated by solar collectors shown in foreground.[78] The flat plate solar collectors and storage tank are located on the ground (instead of on the roof) to facilitate public examination. Solar energy provides hot water, space heating, swimming pool heating, electricity for some lights and appliances and fresh water from liquid wastes. (Photo courtesy of Erich Farber, Univ. Florida.) See page 99.

Solar Water Heater. Water flows through copper tubes soldered in a sinusoidal configuration to a copper sheet (painted black to enhance solar energy absorption) located inside a sheet metal box, 4 ft x 2 ft, covered by a layer of glass. One inch of styrofoam insulation beneath the plate reduces heat loss through the bottom of the box. The 80-gal storage tank is above the top of the absorber so hot water will circulate into the tank without a pump. (Photo courtesy of Erich Farber, Univ. Florida.) See page 83.

Small Solar Still used to reclaim drinking water from household liquid wastes at the University of Florida Solar House. Water evaporates from the metal pan, condenses on the sloping glass cover and is collected as distilled water. (Photo courtesy of Erich Farber, Univ. of Florida.) See page 110.

Thomason SOLARIS home number 6. This home uses a water trickle collector of the type developed by Harry Thomason. Hot water from the collector drains by gravity into a tank in the basement where it is stored for later use for space heating.

Peter Wood Residence in Colorado Springs, Colorado. This home uses the SOLARIS system developed by Harry Thomason.

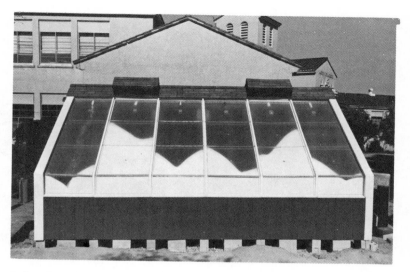

Solar-Heated Greenhouse, double-glassed with fiberglass. At night and on overcast days, polystyrene beads are blown into the 3-in. cavity between the two layers of fiberglass to provide insulation. Greenhouse constructed by Zomeworks Corporation in Albuquerque, New Mexico.

Interior View of Greenhouse, while panels are filling. Dark 55-gal. drums along the north wall are for heat storage. Blower motor units hang from joists in front of heat storage.

National Security and Resources Study Center at Los Alamos, New Mexico. This project was under the direction of Douglas Balcomb of the Los Alamos Scientific Laboratory.

Solar-Heated Office in Mead, Nebraska, designed by James Schoenfelder of Hansen Lind Meyer Architects. South-facing vertical collectors heat air during the winter. Heat storage is in sodium sulfate decahydrate eutectic.

University of Delaware Solar House, built in 1973, uses flat plate collectors incorporating cadmium sulfide solar cells to provide heat and electric power. Lead-acid batteries are used for electrical storage. The house uses dc current for the stove, lights and some appliances, and an invertor provides ac power to the refrigerator, heat pump and fans. (Photo courtesy of Karl Böer.) See page 12.

Experimental Fixed Mirror Concentrator, built by the author and a graduate
student at Georgia Tech, is composed of long narrow flat mirrors arranged on
a concave cylindrical surface. The angles of the mirrors are fixed so that focal
distance is twice the radius of the cylindrical surface. The focus is always
sharp for sunlight of any incident direction. The heat exchanger pipe is pivoted
at the center of the reference cylindrical surface to remain at the focal point
as sun direction changes.[59] See page 37.